Environmental Mana
Plans Demystified

To Catriona, Emily, Janey, Hannah and Oliver

Contents

Preface

> I would like to have a preliminary discussion with an accreditation company to discuss implementing an environmental management system but I don't know what it entails. I have no benchmarks against which to gauge how long the process will take, what it involves, how much it will cost, and whether the information I am given is right for my company.

Listening to this and to the many other frustrated responses of business professionals trying to understanding environmental management and how it may benefit their company prompted me to write a practical guide to understanding and implementing environmental management systems.

This book aims to achieve two objectives. The first is to help you to achieve a comprehensive understanding of the concept of environmental management, enabling conversation with environmental mangers and accreditation companies and enabling you to make informed decisions. Second, if you are the person implementing an environmental management system, this book provides the methods, tools, procedures and documentation you need to detail clearly and explain the environmental management system you are putting into service.

The focus of this book has been on the implementation of ISO 14001. The reason for this is that, at present, ISO 14001 is seen as the leading environmental standard in Europe, and over the next few years it will be the efficiency standard by which ultimately all businesses will be measured.

Stephen Tinsley
June 2001

Acknowledgements

This book is based on the personal experiences, trials and tribulations of the author. However, its construction has involved the skills of many people at Spon Press and many thanks go to them. Special thanks go to Alice Hudson for her guidance and patience, to Jeremy Joseph for his insight and constructive criticism and to Catriona Keir for her excellent illustrations.

Additional thanks go to the British Standards Institution and the Confederation of British Industry for kind permission to use some of their information and to Michael Ramsay of Scotland Electronics (Int.) Limited for permission to use company material for case study applications.

Finally, I would like to thank my family for their tolerance while this book was being written – it could not have been easy!

1 Introduction

For most companies, the business management dynamic is changing. Because of the increasing risk of environmental legislative penalties, the role of environmental management is becoming increasingly important. To be successful, businesses have to follow the environmental management concept and develop product and service offerings that satisfy the environmental needs and wants of consumers more effectively than competitors. Environmental management is a complex task, which incorporates many interrelating parts that interface with consumers, the market place and the actions of competitors. Every company needs an environmental management plan to help examine environmental risk and to ensure that a successful environmental management system is installed to minimise this risk and to create a competitive advantage.

Writing an environmental management plan can be a difficult and time-consuming task if carried out without thought. This book has been written and designed to take the pain out of writing a plan for the implementation of your environmental management system. There are many environmental publications and software packages available today that claim to help and to guide you through the process of writing your environmental management plan. Most of these, however, cloud the issue by trying to dictate rather than to guide. They also assume that future economic, business and environmental trends are known. Although it is important that you take a strategic approach to writing an environmental management plan, it must be flexible enough to react to and exploit new information when it is required. You know your business and products, your customers, your competitors, the business environment in which you operate, and your environmental management objectives. All you want to know is how to write a good environmental management plan; this book will help you to do that.

The book has been written with two main assumptions in mind. The first assumption is that your business is small to medium sized. However, the content of the book will still prove useful if you work within a large business because many departments of larger companies are run like autonomous small businesses and the same principles would still apply.

Second, the explanations or worked examples given are based on the assumption that ISO 14001 is the environmental standard being sought. For those who are not pursuing the ISO 14001 standard, following this book will ensure that implementing lesser environmental management systems will be achieved with ease.

Finally, a good environmental management plan should always include some degree of originality and ingenuity. Therefore, when devising your environmental management plan, you should inject some of your own personality. You may have to convince more senior colleagues or colleagues in other roles or in other departments (for example production managers or sales managers) that your environmental management plan is going to benefit the whole company. It will be you, your knowledge, your experience and your ideas that will capture the interest and imagination of the reader.

Environmental management centres on limiting environmental risk

Current environmental management concentrates on minimising environmental problems. As you will very quickly discover, however, environmental management is really about taking a group of people and setting them new objectives to make the company more efficient and effective. This may take the form of improved operating efficiency or the creation of new product or market opportunities.

The introduction of anything new generally brings its own set of new demands. These new demands require traditional scarce resources, such as time and money, to resolve them. Therefore, environmental management begins with the identification of the environmental factors that affect, or may affect, your business and the prioritisation of those factors to be addressed along with the available resources. You can make environmental management as simple as you want: from writing straightforward procedures for the collection and separation of office paper and plastic to the full-blown manuals and procedural requirements of the international environmental standard ISO 14001.

Let us think for a minute about introducing environmental management into a company. Most managers will have little or no understanding of the concept. They would consider that it does not affect their job or department and that, if it must be introduced, it is someone else's responsibility. To put this into context, focus on the function of the production manager. Production managers do not manage the environment. They manage production all day long, and if they have a stoppage they use, together with a number of well-practised expletives, a portion of their scarce resources to get production moving again. It is quite obvious that the production manager has one overriding objective – to maintain production levels. In most current company cultures, it is unlikely that the environmental issues relating to production activities would achieve even second place on the production manager's list of daily priorities.

From the environmental manager's perspective, the stoppage, or the fixing of the stoppage, may itself have created an environmental issue that might have to be addressed. The environmental manager's resources are also scarce, and resolving an issue in production may leave little or no resources for resolving an issue, for example, in stores. However, environmental management is not about resolving issues as they occur. It would be a bit like the little Dutch boy putting his fingers in the dam to stop the water coming through. The environmental manager, like the little Dutch boy, has limited resources at his disposal (Figure 1.1). You will find that, to be effective, environmental management works better as a system, i.e. it must become part of the whole company's considered business procedures and practices. A simple environmental unit 'bolted on' to existing business procedures and practices will prove ineffective and may be viewed by other managers as a hindrance. The system approach will eventually lead, with time and a bit of luck, to a new company working culture.

Figure 1.1 The limited resources of the environmental manager.

Types of environmental management systems

It may be useful at this point to provide a brief history of the emergence of environmental management systems. Three formal types of environmental management systems have been developed to date. The first system – BS 7750, introduced in March 1992 – was a British Standard for environmental management systems. It has since been withdrawn and largely superseded by the international environmental management system standard ISO 14001, which was introduced in September 1996. The third system, the Eco-Management and Auditing System (EMAS), was introduced in June 1992 but is different from ISO 14001 in that it applies to industrial companies only, is site specific and environmental performance data are required to be disclosed to the public.

At present, ISO 14001 is the preferred option for most companies seeking a recognised environmental management system because it applies to the whole company and environmental data do not need to be disclosed to the public. Although there may be a significant difference in procedural detail between these two systems, their frameworks are basically similar. Both facilitate environmental improvement through the establishment of targets and objectives. The implementation of operational and management procedures ensures that the objectives and targets are achieved.

If a quality system is already in place, such as ISO 9000, the introduction of ISO 14001 will be relatively straightforward. The only significant difference between the two systems is that ISO 14001 has a legislative compliance requirement. This means that a company must be familiar with, and comply with, the environmental legislation and regulations that apply to its business activities. Conversely, if there is not any kind of quality system, the introduction of ISO 14001 first will greatly reduce the time and effort needed to achieve ISO 9000 accreditation in the future. The two systems, in terms of improving the quality of company operations, are viewed by many assessors as being very close in terms of offering a good framework for operating an environmental or quality system.

The multitasking environmental manager

For environmental management to be successfully introduced into a company, the environmental manager needs to be multitasking. The environmental manager needs to wear, at some point, all the hats of all the managers, or persons, responsible for departments or operating functions of the company (Figure 1.2). He or she needs to be aware of how the finance, production, sales, legal and purchasing departments think and operate. The environmental issues that arise will vary from department to department, but they must all be part of the environmental management system.

From time to time throughout the book, you will see a 'winking eye' icon that will signal a hint, a shortcut or a pearl of wisdom from the oyster

Figure 1.2 The environmental manager needs to be able to multitask.

of experience that may save you some inconvenience when implementing your system. When your system implementation has been completed, you may have developed a few more 'hints' of your own that can be passed on.

Environmental managers are not superheroes and cannot leap tall buildings with a single bound. Environmental management will not solve all of your operational or business problems. Placed on the business development continuum, environmental management can be, at one end, a collection of procedures, techniques and attitudes that introduces awareness and efficiency. At the other end of the continuum, however, it has the potential to increase efficiency significantly, reduce operating costs, improve products and services and generate new market and product opportunities. How far you want to go along this continuum is up to you.

The plans, procedures, examples and case studies used in this book have focused mostly on ISO 14001 because it has been assumed that, at some point in the environmental development, you may find there are commercial needs and legislative, ethical or supply chain pressures to achieve this standard.

Detailed below are a few examples of how environmental management affects other departments and how the environmental manager will need to have an awareness of what goes on so that environmental issues in every department are considered in the environmental plan.

Purchasing

This may seem like an uneventful place to begin to consider what environmental issues may be involved. Not so: everything that comes into your company has to be monitored to ensure that it is not an environmental hazard. For example, every substance, such as glue, paint, solder, aerosol cleaners, polishes, needs to be categorised as hazardous or non-hazardous. Each item can be classified by the manufacturer's description on the box or by the information sheet enclosed with the item. If you operate a Control of Substances Hazardous to Health (COSHH) system, then this will already be in place. Those items that are marked as hazardous need to be labelled and securely stored and disposed of, again according to the manufacturer's recommendations.

Marketing

The marketing department is not slow in these days of spin to jump onto a new sale-generating opportunity. Environmental or 'green' marketing is, or can be, a potent weapon if used correctly. Consider for a moment the competitive advantage, particularly with today's environmentally sensitive public, of a product that is environmentally friendly. Many consumers derive a 'feel-good factor' from knowing that once the product has fulfilled its usefulness the packaging and product materials can be recycled and reused. The environmental downside would be the noise pollution created by the constant ringing of the tills as a result of the increased sales!

A close relationship between the marketing manager and the environmental manager allows an exchange of information that can offer additional product benefits, improved company image and increased sales. For example, the environmental manager may suggest the use of different materials to make the product more environmentally friendly, thereby improving the company's image. The marketing manager must determine whether the product price will increase and whether existing consumers will still purchase the changed product.

Product design

In creating the design of, for example, a new car, a team of individuals from a number of disciplines would have grappled with the design rudiments of style, space, ergonomics, aerodynamics, budget and market demand. New issues such as exhaust emission levels, unleaded petrol and recyclability of materials have led to a requirement for the environmental manager to have an input into the product design process to ensure that environmental improvements are included in the design and are not just a cosmetic afterthought.

Production

An effective environmental management system is an ideal tool with which to monitor the environmental issues involved in changing a product design or installing a new production process. All the environmental implications of production changes need to be mapped, assessed and monitored as part of the new environmental management system.

Public relations

For most companies, good public relations (PR) are everything. Bad PR hurts business. Bad environmental PR is disastrous. Good environmental PR is excellent for business. The environmental manager and the PR manager need a close working relationship to ensure that good environmental developments and endeavours reach a wide audience. A strong environmental public image is a cherished prize – but beware of false promises. Early consumer product manufacturers were too quick to jump onto the green-product bandwagon. Consumers identified too quickly those companies promoting the same product with a green promise, and what followed for these companies was a serious loss of credibility and market share.

Administration

The following is a conversation recently overheard between the environmental manager (Tom) and the accounting/administration manager (Dave) in a medium-sized company within the construction industry.

Location: staff kitchen (Figure 1.3)

Dave: Morning Tom, may I have a quick word about this questionnaire you have sent out regarding the new environmental management system currently being installed?

Tom: Morning. Yes, certainly. What would you like to know? Can we get a coffee before we start?

Dave: I really don't see the accounting and administration department having any contribution to make, or even having any environmental effect at all. Shouldn't this be directed to engineering and production?

Tom: Do you see those two cardboard boxes there in the corner, one labelled 'plastic' and the other labelled 'paper'?

Dave: Yes.

Tom: Put your coffee cup in the box marked 'plastic'.

Dave: OK.

Tom: You've just contributed to the environmental management system.

Figure 1.3 Contributing to the environmental management system.

It can be unsettling to know that everyone can have a different interpretation on how much or how little they affect the environment. Whether the contribution is big or small, the environmental management system should pervade all departments and functions of a company. An environmental manager will find him/herself communicating, at various levels of understanding, with all the managers and other key persons within the company to ensure that there is a constant transfer of information to aid the effective implementation and operation of an environmental management system.

As it currently stands, ISO 14001 is the gold standard of environmental management achievement. That said, it is not necessary to implement the full international standard ISO 14001 overnight. A simple system can be introduced initially, with a further staged introduction applied over 1–2 years, eventually achieving full ISO 14001. Whatever speed you wish to progress at and whatever route you choose, this book will help you to achieve your goal.

2   Managing the environment

In this chapter, advice is given on the terminology used in environmental management. It is worth bearing in mind that 'environmentalese' should be kept to a minimum when communicating your environmental plans and objectives to others. Remember, this is a new language and many people will be unsure of what is being said and are likely to ignore the subject so as not to demonstrate a lack of knowledge. Communication is key to environmental management, so it is important to use a language that everyone understands to get the message across. When conversations do centre on environmental issues, however, the discussions within this chapter will assist with environmental management understanding.

Environmental management is very much in vogue at present, and its star is rising. Many managers and the general public are aware of the term environmental management but are unsure of its use, benefits or terminology. In a fast-moving business environment, where do managers get the time to learn this new language and adopt new techniques? It is another form of management strategy, after all, to improve efficiency, so why is there a lack of general knowledge to aid its introduction into organisations? One reason may be a lack of environmental training, informative literature and knowledge within the ranks of middle and senior managers (Figure 2.1).

A survey undertaken by the Confederation of British Industry (CBI) in 1992 demonstrated the lack of environmental awareness among senior managers. Figure 2.1 shows the high level of environmental awareness in younger people. This bodes well for those people moving into future managerial positions. The current position, however, is that middle and senior managers with little environmental awareness are still making the majority of business decisions. It is accepted that some middle managers, and possibly one or two senior managers, have sought additional environmental training or have chosen an environmental career change, but they seem to be still in the minority.

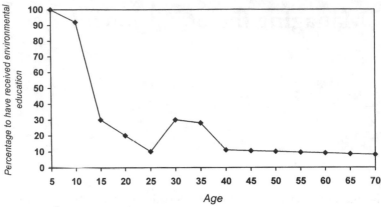

Figure 2.1 Environmental education age profile.

Introduction to environmental management

The introduction of environmental management always takes the form of a system. Why a system? Because you need more than a quick fix environmental 'elastoplast' stuck onto an operational sore to prevent infection. It is the cause of the issue that has to be treated and not the symptoms. A system needs to be put into place because it has to become part of the company ethos to ensure that every activity, operation or process is managed for environmental improvement. The introduction of your environmental management system should have the following character-istics:

- a clear objective;
- a fixed time-scale;
- a team of people.

A clear objective

Let us be pragmatic about environmental management systems right from the start. If you are the person responsible for the implementation of an environmental management system, you will not greet your fellow employees with a big 'S' on your shirt and red underpants on over your trousers (Figure 2.2). You will not install the system into the company to save endangered species or reduce global warming. You might install it to protect your company from environmental legislation, to reduce operating costs and/or to improve efficiency. Whatever your objective or objectives may be, make it/them clear to yourself and everyone else at the start in a clearly written one-page environmental policy statement. You will not be looking to save the planet; you may, however, look for gradual

Figure 2.2 The environmental management 'superhero'.

improvement in the way your business activities affect the environment, and you will want to be more competitive and more efficient.

A fixed time-scale

Whether installing a simple environmental management system over a few months or installing ISO 14001 over 1–2 years, specify start and completion dates, and let everyone know what these are; also, before you start to undertake this mammoth task, it would be advisable to set interim dates for the achievement of specific tasks. It is true that some projects within a company seem to emerge from the ether and disappear back there with very few employees being any the wiser. Introducing an environmental management system should be considered a deliverable that has to be delivered.

The introduction of an environmental management system will be viewed by many in the company as unnecessary and of no practical operational benefit, and many intentional and unintentional barriers will be put in its way. The setting of key dates ensures that environmental deliverables are clearly completed on time. Make the deliverables and the completion dates known throughout the company; it will help to focus everyone's mind on the job at hand.

 **Hint**

Although managers and employees may seem uninterested and slow to accept environmental change, they will be watching your commitment to the project very carefully. If you are not committed to the project or you let deadlines slip, they will not be slow in letting you know that fact, particularly when you are asking them to meet deadlines and to demonstrate commitment.

A team of people

You may have heard of the expression 'too many chiefs and not enough Indians'. When it comes to forming an environmental working group (EWG), have as many Indians as is practical but ensure that you have at least one chief, the bigger the better (Figure 2.3). A cross-disciplinary selection of Indians is very helpful, and they will certainly do the majority of the organising, writing, planning and various other bits to ensure that an environmental management system is installed. The chief, however, will be required to smoke many political peace pipes in many senior wigwams to ensure the success of the system.

The EWG should be tasked with meeting frequently, if appropriate at least once a week, in order to discuss any matters that have arisen and to ensure that steady progress is maintained. It is important to log in your diary that the meeting has been held, and that action points are noted and tasked to an individual with a date for completion. Periodically, a memorandum or newsletter should be released to advise everyone of changes and progress. If they can be held, company gatherings of all staff can put the message across quickly and provide the opportunity if not for bonding then at least for a question-and-answer session. With a little prior warning, you could take advantage of the briefing opportunity and use it for some internal environmental management training.

A change of culture

When introducing an environmental management system into a company, you will find that the more committed you and your company become to addressing environmental issues, the more you will understand the extent to which products, services and processes must change. You will also become aware that merely setting new targets and objectives cannot bring about the scale and scope of the changes needed. What is required is a complete cultural change.

Figure 2.3 The big chief.

All good business management textbooks will tell you that one of the functions of a manager is to provide stability to a company. Managers are tasked by the owners of a company to ensure that a certain level of income and profitability is achieved with the minimum of company disruption. This philosophy does not apply to the environmental manager. The role of the environmental manager is to effect change. While other managers are managing stability, the environmental manager will be aware that, when introducing environmental management and change, the support of these other managers will be needed.

Environmental managers will metaphorically be slapped in the face with a cold kipper when asking production managers to change product processes for environmental improvement. The environmental manager will be advised in a very unambiguous manner that daily production activities take precedence over low-priority, long-term environmental objectives. Therefore, the role of the environmental manager is to bring about change with the help of those managers that are trying to maintain company stability. It is for this reason that high-level support and broad buy-in are needed at the start.

What is environmental management?

Everything we do has a greater or lesser impact upon the environment. It is generally accepted that there are finite amounts of resources on the planet – there is only so much oil that can be taken from the ground. Everyone is aware that many small things can be done to make better use of resources but that we tend to leave it to other people to do this, and consequently very little happens. Current environmental thinking suggests that a major improvement can be made if industry manages the resources that it uses more efficiently. With this in mind, the European Union (EU) has gradually sanctioned an increase in environmental legislation that penalises industrial bad practice or inefficiency.

As environmental legislation continues to increase, companies are learning to manage rather than to battle or avoid environmental risk. The introduction of environmental management systems into companies began with BS 7750, the UK's environmental standard, which followed the basic guidelines of the old BS 5750 quality standard. Europe followed this by creating an International Standard, ISO 14001. The two standards were based on a similar structure, and, over time, BS 7750 merged into the main environmental standard of today, ISO 14001.

The multinationals were the first to introduce ISO 14001 into their plans. Those companies that went through the process of implementing an environmental management system are now witnessing benefits other than legislative compliance. Many large organisations have identified numerous company drivers for the introduction of environmental management systems and are pressurising their suppliers to adopt the same environmental attitudes.

Why ISO 14001?

Having determined what environmental management is, the next question to answer is 'Why ISO 14001?'. There are other management systems that can improve operational efficiency and create new business opportunities; ISO 9000, the ISO 9001 standard, is one obvious example. None of these other systems, however, protects against environmental legislation. ISO 14001 requires the monitoring of environmental legislation and the assessment of its risk to a company. Fifteen years ago, there was only a small amount of environmental legislation that posed any kind of threat to a company's operation. Today, there are over 300 articles of UK and European legislation that companies need to be aware of, and the regional environmental agencies have much greater policing powers over business activities.

'Ignorance of the law is no excuse' is a well-worn expression. You may well have witnessed, or even been party to, the burning of office waste paper in a 45-gallon, or 205-litre, drum at the rear of a company premises.

An innocent enough pastime you may think. Such an action today brings a maximum fine of £20,000 to both the unknowing and the uncaring. The number of environmental laws is still growing, and in another 10 years who knows how many more will be on the statute books.

Key company drivers for environmental management system (EMS) application

Minimising environmental risk exposure is just one of the key drivers to companies adopting ISO 14001. Other drivers include:

- energy efficiency;
- waste minimisation;
- green company image;
- competitive advantage.

Environmental risk minimisation

Over the past 15 years, environmental legislation has grown from consisting of a few items to consisting of hundreds. The penalties for companies transgressing these acts have also grown to incorporate significant fines and imprisonment for company directors. Those companies that have polluted in the past can no longer escape their clean-up responsibilities. The 'polluter pays' principle is central to existing environmental legislation and ensures that pollution caused by companies in the past is still their responsibility today.

The more attention that you give to environmental management issues the more likely you are to reduce your company's exposure to environmental risk and the potential to incur penalties under legislation governing waste, emissions, chemical storage and related health and safety issues.

The majority of large companies have environmental policy statements declaring that their objective is to improve environmental management and reduce exposure to environmental risk. Despite growing environmental pressures, however, most managers still hold to the notion that pollution pays and that pollution prevention does not.

Most managers who believe that the pursuit of environmentally sound strategies is detrimental to the principal managerial goals of profitability, maintaining market share, cost control and production efficiency fuel the persistence of this view, which rests upon three fundamental assumptions:

- The benefits of following environmentally sound practices cannot be achieved because consumers are not prepared to pay the increased costs to industry.
- Business costs of environmentally sound strategies are high.

- Industry is not prepared to increase investment in non-productive areas such as environmental risk aversion.

Over the past two decades, these strongly held tenets of traditional management thinking have begun to break down slowly. Managers are beginning to see the benefits of improved efficiency and competitive advantage that come from addressing environmental issues. The main objective of economic policy is to maintain and gradually improve the standard of living for the existing population and for future generations; this is also the main objective of environmental policy.

This main objective of environmental policy will be achieved by avoiding or limiting the problems with pollution, with dereliction and with loss of habitats and wildlife species arising from human activity. Short-term conflict can arise when the cost of avoiding environmental damage is perceived as being a constraint on economic activity; in the long term, this conflict should be overcome.

By addressing growing public concern for environmental sensitivity and the legislative requirements of industry, companies are gradually becoming more aware of the benefits of having an environmental focus. Research has shown that there is profit in going green and that sound environmental management can lead to competitive advantage in business practice.

Despite the positive business development aspects presented to industry by current environmental management research, there remain studies that show that not all environmental management systems are accepted in companies. Owing to a lack of environmental awareness and basic operational misunderstandings, managers often fail to identify whether environmental management is a functional or corporate activity. As a consequence, environmental management initiatives have floundered as a result of lack of company and management acceptance and commitment.

Despite assurances from bodies such as the Confederation of British Industry (CBI), the Department for Education and Employment (DfEE) and the Department for the Environment, Transport and the Regions (DETR), industry has been slow to accept the argument that there is profit in going green. Companies are being shown the advantages of environmental business opportunities by these bodies developing programmes of waste reduction, material reuse and recycling and promoting energy efficiency by introducing environmental management systems.

An example of this type of support comes from the Department of Trade and Industry (DTI). It has released a number of publications on waste reduction and energy saving for companies, particularly small- and medium-sized enterprises (SMEs). The DTI, together with the British Standards Institution (BSI), is currently undertaking a pilot study to simplify the introduction of ISO 14001 into SMEs.

Energy efficiency

This is the most logical starting point for those companies wishing to begin with something familiar that will provide a short-term return for minimum expenditure. A simple review of oil, electricity and gas bills will provide a base from which future savings can be measured (see Box 2.1). Adopting an energy efficiency programme is a good way to begin an environmental awareness programme for a company. Put very simply, an environmental policy will state that the company will use energy more efficiently. One company objective might be to reduce energy consumption by 10 per cent, measured against existing energy bills, in the first year.

Waste minimisation

Many companies expend much of their business development budget improving production or increasing sales. Company management will find that there is a greater return on investment, as high as 10 per cent of turnover, if the same importance or investment is attached to improving waste management. Reducing waste improves profitability, and any savings go straight to the bottom line and improve competitiveness. If a company is not ready to pursue the ISO 14001 environmental standard at present, then an overall attempt at a waste minimisation programme will improve business efficiency and reduce environmental impacts in the short term. Simple design changes in a product or processes could result in fewer natural resources being used to convert raw materials into the final product.

Green company image

Businesses strive continually to be different from their competitors, and, in an attempt to gain an edge, products or services are linked to environmental benefits. Many industry sectors are becoming increasingly aware that businesses and the general public prefer, where possible, to deal with companies that are able to demonstrate a willingness to operate in an environmentally responsible way. Take, for example, the Body Shop; Anita Roddick, its founder, saw at a very early stage the benefits to her company if it was perceived to be environmentally friendly. Body Shop's green image went further: by adopting many environmentally friendly policies, some of which were controversial, customers were of the opinion that the company had sound ethics – it was perceived to be honest and forthright in all business matters.

Competitive advantage

If a company improves its efficiency in its use of resources, particularly in its production processes and use of energy and water, it will gain an advantage over competitors that remain inefficient. Internally, efficient

Box 2.1 More efficient use of energy

Energy efficiency survey

Local authorities and Local Enterprise Companies (LECs/TECs) often provide free or heavily subsidised energy efficiency surveys for companies within their area. Some of the findings are simple in nature and cost little to implement, and yet can achieve significant savings.

A typical programme of energy savings identification will begin with an examination of how the company pays for electricity, gas, oil, water and effluent currently to ensure that it is buying them at the best possible prices. Then, an assessment of company operations will be undertaken to determine potential methods of waste minimisation and energy usage maximisation.

A simple energy efficiency survey on a small company with approximately forty staff would result in a saving of £2,500 per annum. The breakdown of potential savings can be summarised in Table 2.1.

Energy savings analysis

The energy savings analysis demonstrates that, with no investment, significant savings can be made either by renegotiating the rates with the company's current electricity supplier or by changing supplier. This is a no-investment strategy and pays dividends, creating a potential saving of £1,500 (see Table 2.1). A similar approach should be undertaken with other energy suppliers.

A low-investment strategy would involve the purchase of heating thermostats or optimisers to switch the heating on and off depending on external weather and seasonal conditions. Slim-line fluorescent tubes use approximately 8 per cent less electricity and cost the same as standard fluorescent tubes. A little more expensive, compact fluorescent tubes use 75 per cent less electricity, last eight times longer and lead to an obvious reduction in replacement and maintenance costs. Using this strategy should realise an additional £200 of savings (see Table 2.1).

Investing in more efficient plant and equipment can involve significant capital expenditure and, as a result, can constitute a high-investment strategy. New heating systems will be more efficient and, if planned correctly, the initial capital expenditure will ensure future energy efficiency, low maintenance and eventual cost savings. They will also conform to the latest environmental legislative requirements.

Table 2.1 Energy savings analysis

Investment strategy	Annual savings (£)	Savings (%)
No investment	1500	27
Low investment	200	3
High investment	800+	14
Total saving	2500	44

heating and lighting systems and safe handling of hazardous substances result in greater profitability, improved working conditions and a boost to staff morale.

An environmental management system

What then is an environmental management system? How does it work?

Taking the formal approach first, an environmental management system (EMS) is defined by the British Standards Institution as: 'the organisational structure, planning activities, responsibilities, practices, procedures, processes and resources for developing, implementing, achieving, reviewing and maintaining the environmental policy.'

You are not alone if you did not fully undertand this statement when you read it the first time. So, a more user-friendly description is offered to support the above. It goes like this: an environmental management system is 'a framework for implementing your environmental policy and objectives and, through recorded evidence, achieving conformance with the standard that can be demonstrated to others.' Environmental management systems will always differ significantly in procedural detail. Nevertheless, they will share a similar framework and approach for environmental improvement.

To introduce an environmental management system – let us say that it is set up according to ISO 14001 – the first thing is to establish broad environmental goals (environmental policy statement). Then, the company's activities and processes should be assessed (environmental audit), followed by the environmental implications of its business operations (aspects and impacts analysis). From this information, it is possible to set environmental improvement standards (objectives and targets) and establish a programme to achieve the objectives and targets (operational and management procedures). When all of this is in place, the progress made in achieving the objectives and targets (environmental management system audit) may be evaluated and, finally, the system in the light of the findings (management review) can be reviewed.

Figure 2.4 demonstrates the EMS process. The underlying emphasis of the system is on continuous improvement; this can occur in two ways. First, as objectives and targets are set and achieved, so others take their place and the company continually makes its operations more environmentally friendly. Second, the environmental management system itself improves continuously by becoming more streamlined, e.g. initial procedures may seem cumbersome with unnecessary paperwork for the size of the company. Regular EMS review meetings with employees will assist with the streamlining and make the EMS fit the way a company operates.

Having bracketed some of the key actions of the environmental management system above, something that is familiar and perhaps a little

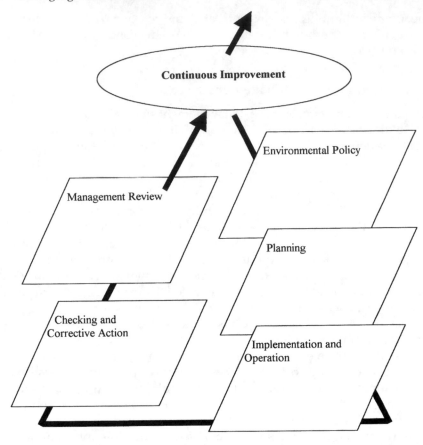

Figure 2.4 EMS continuous improvement model.

comforting may be obvious. Many of the steps of setting goals, objectives and targets and evaluation and review are the basic elements of quality management and health and safety. If this system framework is already in place, use it.

After a recent seminar on introducing ISO 14001 into SMEs, a gentleman stood up during the question-and-answer session and stated that he ran a company with twenty-seven employees. He said that the company had ISO 9001 in place and that he was adding ISO 14001, which would create a duplication of systems not to mention a duplication of paperwork. The answer to this comment is very simple: an environmental management system should be incorporated into a quality system or vice versa. The aim should be to create an integrated quality, health and safety and environmental management system. If management systems start out as separate units, don't be too concerned, they will eventually – to paraphrase the Spice Girls – become one!

To be totally pragmatic about the whole exercise, introducing a formal environmental management system, for example ISO 14001, does not require levels of environmental performance to be specified. What is required is that such systems comply with current legislation, meet the terms of the standard and follow procedures for continuous improvement for environmental performance. The hint given here is not for a company to generate shortcuts to the system, but rather for it to be aware of the built-in flexibility that will allow it to make steady environmental progress to its long-term competitive and economic benefit.

The range of complexity of environmental management systems ranges from the basic environmental audit to the sophisticated sustainable development strategy (see Appendix 1 for more detail). As some corporate change is required with the implementation of any strategic change, the more radical the change the more resistance will be encountered and the more resources will be needed to overcome this resistance. To reduce the level of resistance, it is recommended that companies adopt a phased approach to introducing environmental management systems based on the resources available and the extent of commitment of senior management.

Those companies with existing total quality management (TQM) systems saw the benefits of decreasing manufacturing costs, faster time to market and increased market share that would follow by introducing environmental management systems. However, while some companies were achieving success with environmental management strategies, others found their strategies had to be shelved or abandoned altogether. The main reasons for such actions were lack of support for the environmental management system by middle managers, lack of commitment by senior management and inability to communicate the benefits of further environmental investment. Because of the increasing severity of penalties for environmental management failure, large organisations have adopted environmental management systems that focus on the 'management' of environmental risk.

Why you need an environmental management plan

An environmental management plan can be used to identify new business opportunities for a company as well as to show how it intends to achieve environmental management targets and objectives by satisfying the environmental needs and wants of consumers. As a communication tool, an environmental management plan can be used to integrate all the environmental management activities of a company into one co-ordinated and easy to communicate action plan. Environmental management plans help managers to make explicit environmental management decisions within a dynamic organisational environment, through a systematic process to achieve position, survival, growth and sustained competitive advantage within specified time horizons and acknowledged resource constraints.

A good environmental management plan will be useful for a number of reasons. It should:

- decrease environmental risk and lead to increased control over a company's future;
- provide direction and guidelines for the introduction of new products;
- help to make changes to the design of products and processes;
- contain enough detail of a company's environmental management objectives, activities and direction to allow the day-to-day implementation of environmental management activities to be carried out by junior employees or managers;
- give all staff a better understanding of the environmental management activities and the importance of environmental management to the company;
- encourage company identity and team spirit and increase the motivation of both environmental management and non-environmental management staff associated with the planning process;
- help a company to keep ahead of its competition.

Environmental management plans are usually highly confidential and therefore read only by staff members in a business. However, in some cases, environmental management plans may be useful when applying for funding from environmentally sensitive financial institutions. The Co-operative Bank prides itself on supporting companies that are environmentally aware, and the presentation of an environmental management plan, together with the obligatory business plan, may provide the bank with greater investment confidence.

What is environmental planning?

For most, this is a journey into the unknown, so it is important to have a good plan and some clear landmarks by which to navigate. Consider yourself a pioneer environmental manager. It will be your responsibility to be a guide and source of inspiration if you are the one tasked with preparing an environmental plan for the introduction of an environmental management system. Why pioneer? Because you will be exploring the environmental unknown of the company and management, and employees will be relying on you to talk constantly to them and explain strange sights and sounds and dispel any rumours of ghosts or monsters. You will need to chart your progress carefully and to provide clear measures of how far everyone has come and what has been achieved.

A pioneering environmental manager will be constantly thinking ahead about deadlines that need to be met. Going for ISO 14001 means that there will be clear deadlines with which to comply. Fortunately, negotiating with your assessment company can set these deadlines. It is possible to set tight or relaxed deadlines depending on business requirements at the time. Remember, if you are the environmental manager, this may not be a full-time working commitment; you may wear another corporate hat, possibly a conflicting one, so build in some slack.

For the keen environmental manager, there are two main elements to implementing an environmental management system into a company. And, staying with the explorers theme, they are identified as follows.

Navigation

You may find, initially, that installing an environmental management system seems to create more problems than it solves. This is true. But these problems are small, and going through the plan that you have drawn up for yourself will ensure that these problems remain small and are overcome as the process continues.

Figure 2.5 It is important to keep checking the signs.

Constantly monitoring the situation helps. For example, an advanced driving instructor was relaying a frustration about a pupil who would not look short and long while driving. The driving instructor advised his pupils to vary their eye-line when driving in a built-up area. As a ready rule of thumb to improve driver awareness, he encouraged them to change their eye-line continually by looking first 150 metres ahead, then 100 metres ahead, then 50 metres ahead, back to 100 metres, then 150 metres and so on.

The same technique, used to improve driver awareness, works well when navigating the potential pitfalls for an environmental management system. To ensure that the system continues on the right road, keep checking the plan and the relevant deadlines. Continue checking the signs for what has to be done now and what has to be done next and be aware of how these actions tie in with meeting future deadlines (Figure 2.5).

Charting progress

Support for your journey is critical. You cannot afford to lose anyone *en route*. Keep everyone together with well-signposted routes. One of the benefits of following an accredited environmental management system is that deadlines, once agreed, have to be met. The deadlines agreed are phased assessment deadlines. You may, for example, have agreed three interim assessment dates. This is the assessor's method of charting progress, but it is also a sound method for ensuring that you are building your environmental management system on a solid foundation.

3 Nine steps to a successful environmental management plan

This chapter takes the environmental manager through a series of steps essential to delivering a successful environmental management plan. It will provide an insight into the true framework of a successful environmental management plan and the reader will quickly become acquainted with many potential, and easily avoided, pitfalls. Although the following chapters will provide a framework for the introduction of the technical aspects of an environmental management system, this chapter will provide some awareness of the preparatory activities that can assist with the implementation of the system.

If you are the person responsible for ensuring the successful implementation of an environmental management system, following these nine steps will provide some stress relief. Whether you are installing a simple waste management or energy efficiency system or a more complex environmental management system, such as ISO 14001, the nine steps will provide a framework for the preparation of a successful implementation process. The assumption made when compiling the nine steps outlined below is that your company has decided to take the plunge and implement the ISO 14001 environmental standard.

The nine steps

1   Define what has to be done.
2   Form the basis for the plan.
3   Obtain management agreement.
4   Assemble a working group.
5   Seek employee feedback.
6   Set deadlines.
7   Communicate.
8   Provide visible signs of progress.
9   See people as shades of green.

Define what has to be done

Whether the company is big or small, it is worthwhile defining what the environmental management system is there to do, i.e. whether it is to improve efficiency or to project a green image. The detail of these initial objectives will be incorporated into a corporate policy statement that will form part of the environmental management plan. As the person responsible, you may also want to determine the length of the project, who else will be involved and what your particular role is likely to be. At this stage, you will probably identify general key activities that need to be undertaken to start the implementation process.

To determine where you are going and how you are going to get there, it is useful to understand where you are now. An environmental audit of your current business activities is an important next stage. The intricacies of the environmental audit are explained more fully in Chapter 4; at this point, a brief overview of its content will give a flavour of what would be included in the audit.

Any environmental management audit will involve an analysis of both external and internal factors. External factors can be divided into external environmental factors and customer and competitive factors. External environmental factors relate to circumstances that an organisation cannot directly control and include broad issues such as the economic, customer, competitor and operational policies.

The audit must also encompass a breakdown of internal factors, and analyse the resources and skills of an organisation, including any deficiencies or weaknesses. Finally, consideration should be given to the more immediate environmental factors of customers' needs and competitor activity that environmental management strategies may be able to influence or affect to some extent.

Form the basis for the plan

The environmental audit allows the environmental manager to set a strong foundation for the overall environmental management plan. The plan, like a good suit, should fit the 'cut' of a company. Once the senior management directive has been received that an environmental management system is to be installed, the plan will not just appear as a flash of inspiration. Therefore, the next step is to determine the environmental issues that affect the company. The audit will identify all the factors of a business's activities that will have an impact upon the environment. Once these factors have been prioritised, the environmental targets and objectives for the company can be set.

A corporate environmental plan is the basis for the whole environmental management system. The plan created should eventually reflect a business's activities and the impact that it has upon the environment. If you are the person responsible for the implementation of the plan, do not fall into the trap of an electronics manager who, when asked what impact his company had on the environment, replied 'we are in a clean industry, we do not manufacture anything, we do not have any impact on the environment.' Everything a company does, whether it supplies products or services, affects the environment to a greater or lesser extent.

Obtain management agreement

The basis of the plan has been outlined, what next? Unfortunately, at this moment, it has the support of only one person – the person responsible for implementing the plan. If you are that person, you now have to secure managerial support for the plan. If you are not intending to obtain managerial support for the implementation of ISO 14001, stop reading now and throw this book into the nearest bucket. You need managerial support for this exercise.

It is not even possible to play a political numbers game with the plan either. If there are ten managers in the company or strategic business unit, securing the support of six does not necessarily provide any significant advantage. The reason for this is that an environmental management system will pervade all functions of the company and affect every person and therefore it will need full management support to ensure company-wide acceptance. Of course, it does not have to, and is unlikely to, gain 100 per cent support, but the plan needs enough support to ensure that all managers will implement procedures.

There is a need to reassure individual managers of the benefits of the system in order to win their support. As the implementer of the plan, spend time with each manager explaining what is to happen, what procedures are being implemented and what they mean to that manager's operation. Help the manager to a full understanding of what is happening, not just in his or her department but also in other areas of the company. Stress the benefits that will come about as a result of the new procedures, and offer the manager your support and that of your working group in answering any questions or assisting with any issue that may arise. Offer training sessions if there is a need, or agree to them if they are asked for.

Assemble a working group

A major outcome of the initial environmental review is the formation of an environmental working group (EWG) headed by a director or senior manager. The EWG should be given the task of meeting periodically – the frequency will depend upon the company size and the size of the scheme – in order to discuss any matters that have arisen and to check the progress of any on-going environmental topics.

The people chosen for this group do not need to be environmental management experts, but they do need to be committed to, or convinced of, the potential benefits to the company and their own working environment.

The EWG should consist of a company director, or senior manager, and at least two co-opted members of staff, and should be responsible for:

- introducing an environmental management system into the company;
- co-ordinating environmental activities within the company;
- acting as a focal point for environmental issues;
- producing environmental progress memorandums or newsletters for circulation within the company;
- seeking employee feedback;
- environmental aspects and impacts;
- development of an environmental policy statement.

To aid you, as implementer, in your quest for certification, it would help progress if as part of your team you had a technical author, or someone with good writing skills, and an engineer or two. The writer will prove to be an invaluable member of your team. Half of the installation process is focused on procedures, documents, manuals, forms and other pieces of necessary paperwork. If this paperwork is prepared when it is required, the whole process will speed along. There will be many changes to the written material as implementation progresses, and someone skilled and dedicated to record the changes, cross-reference them and revise the relevant pieces of paper will be needed. As an additional benefit, the writer will become very familiar with the content of the environmental management system and will naturally become a useful person for creating and conducting the auditing procedure for the environmental management system.

If the company is engineering based, the engineering members of the team will also prove to be valuable allies in the battle for acceptance and could provide two main benefits. First, their engineering knowledge will assist the writer in compiling environmental operating procedures for products and processes, and, second, their involvement will bring the engineering element of the company on board more easily. For those in other industry sectors, the team should have representatives from all key departments (Figure 3.1).

Seek employee feedback

Employee involvement in environmental matters is fundamental to the adoption and maintenance of a successful EMS. It is unlikely that you, as implementer, will have all of the environmental management answers at the outset. Even if you do, it is still politic to ask other members of the company for their input. You will be amazed at how knowledgeable most employees are about environmental issues and how they affect the working

Figure 3.1 The environmental management team should have representatives from all key departments.

environment. A company-wide or departmental questionnaire can produce a number of suggestions as to how employees could be involved in the EMS. If the questions are pertinent, they can also provide suggestions as to how a company's operations could be improved in terms of environmental protection. All employees should be urged to highlight any process or operation that could have a significant environmental impact. Every operation has an impact, big or small, on the environment. The calculations for determining the degree of impact are given in Chapter 4.

The responses to a questionnaire will be many and varied. Some will be positive, others negative; all need to be considered. The responses, even the most obscure, should be treated professionally. They represent raw survey data about a company and its activities, and employees' perceptions of how these affect the working environment. Whether or not it is relevant to the environmental management system , it is still useful information and should be treated as such. These data should be recorded, analysed and prioritised as they will provide the basis for setting the objectives and targets for a corporate environmental plan. When the employee responses have been listed, those that may apply to health and safety, quality or other systems need to be separated. The items that remain and that are relevant to the EMS will need to be assessed for their significance and whether they need to be included as objectives or targets (see Chapter 5).

Set deadlines

It may seem to you – as implementer – when looking at other company activities, that deadlines are set and are then changed or slip without too much fuss being created. Let deadlines slip when introducing an environmental management system and you will receive little understanding or sympathy from your colleagues. If, for example, you are introducing ISO 14001 into your company, you will be set, and have to meet, strict deadlines by your assessor. These will be phased over a number of months, and each will rely on the previous deadline being met before progress to the next stage can be achieved. Let one slip and they all slip, and you may find that money, time and resources have been wasted and the implementation process could stall or fail altogether.

The setting of assessment stages over several months can lull the working group and employees into a false perception that there is adequate time for completion of the process. To ensure that time is not squandered or lost, set your working group weekly targets. If you have frequent meetings, and it is strongly advised that you do, you should ensure not only that frequent targets are set, but also that members of the working group are tasked with their completion.

Frequent meetings will keep the project in the company spotlight as employees will enquire continually about the progress being made and working group members will be reporting progress back to their own managers or departments. Another benefit of frequent meetings is that employees will be aware that any actions that require certain targets to be achieved will be in the spotlight; a positive and speedy response is the most likely outcome.

Communicate

Take every opportunity that you can to communicate what has been achieved, what is happening and what you are planning to do. Start as you mean to go on. At the very outset, as you are distributing the environmental questionnaire, attach a memorandum informing all employees about the reason for the survey and advise them of the expected completion date. Take the opportunity to introduce the environmental working group and identify areas where it may be able to assist.

The chances are that few will listen at first, that some will forget and that others will be on holiday. As every good advertising manager will tell you, an advertisement, no matter how creative, needs to be exposed to the targeted purchaser for a minimum of five times before action is taken and the product or service is purchased. You may even find that this small piece of wisdom is a little optimistic when trying to get everyone to accept your new environmental management system.

Essentially, to maintain steady progress and ensure that project momentum does not stall, frequent progress updates and target setting are a must. It will happen that meetings are missed by some or even postponed altogether, but having regular meetings for 15–30 minutes will keep the process going and ensure that any missed meetings will result in a minimum of lost information and / or delayed targets. Where possible, a full company or departmental turnout for an overall briefing and question-and-answer session at least once each month or quarter should be arranged. This provides a strong message to everyone that communication is happening.

Every opportunity to maximise communication opportunities should be taken. Large briefing sessions involving all company employees offer the opportunity of airing any problems that may be developing and of airing issues that have been overlooked or are newly created. These group sessions can also double up as internal environmental training sessions that will be of benefit later in the implementation process. Whenever there is an opportunity for environmental communications, even a small one, you should still try to maximise it as these opportunities will all help to impress the assessor at the end of the day.

Provide visible signs of progress

As you and your team are busying yourselves with writing additional procedures, creating waste logs and keeping everyone informed through additional memorandums, you will need to heed people who comment on the anomaly of creating more paperwork by introducing an environmental management system. This will happen from the outset of the project. When you have read this chapter, however, you will have had the necessary notice to come up with a witty reply to the inevitable question 'How many trees have you cut down to introduce this environmental management system?' as you hand out another memorandum.

In the age of electronic communication, it is worth pointing out that not everyone reads his or her e-mails. The reading and actioning of environmental management e-mail may not be high on everyone's 'do today' list. If practicable, the personal distribution of paperwork, memorandums, etc. is preferable to e-mails as there is the added advantage of explaining, convincing and recruiting support for the new system when handing them round.

Before a company signs up for the ISO 14001 standard, an accreditation company will make a preassessment visit. The main reason for the visit is to determine how much of the environmental standard a company already complies with and how much is still to be achieved. The company will be fortunate, unless some environmental issues have been addressed in the past, to come away with many ticked boxes. That is acceptable and very normal, and as implementer of the EMS you will receive a clear indication of how much work there is to be done. Do not be put off by the lack of ticks; you will be amazed at how much progress can be made even after the first assessment. Following this preassessment visit, you need to make it known to everyone, by way of a circulated report or copy of the completed assessor's questionnaire, how much has to be done. If you are going to transform a jungle of a garden into a manicured lawn, you should take a picture before you start so that you can show everyone the transformation after you have finished (Figure 3.2).

See people as shades of green

Before you start telling people that you are introducing an environmental management system into the company, take a moment and think how they may perceive what you are trying to do. Some may think that you are out to save the world, starting of course with the company. Some may see it as another cost-saving initiative to increase company profits. Others will have thoughts somewhere in between these extremes, all of which may vary in their degrees of environmental awareness. You will, hopefully, change most of these perceptions over time. At the beginning of the process, it is as well to assume, and accept, that all sets of employees will have a range of environmental awareness. It would not be too flippant to suggest that:

> ... Environmental
> awareness exists
> in shades
> of green

Figure 3.2 Taking a 'picture' of a company after the preassessment visit can make it easier to see how much progress is made.

You may wish to classify those that have no interest in environmental issues as 'browns'. Those with some awareness you may classify as 'light greens', whereas those employees that have good environmental knowledge may seem to be 'dark green'.

Five possible classifications have been listed below as a guide. The reason for the classification is to make the point that there will be few, if any, dark greens in your company unless you work for The Body Shop or Ben and Jerry's. It is most likely that there will be a mixture of browns, light greens and passive greens. This suggests that you need to have good leadership skills to implement your environmental management system. Usually, however, the majority of people are 'for' the system, even though their levels of contribution will vary, and very few are actually against it.

- Brown – no knowledge or interest in environmental issues.
- Light green – mildly aware.
- Passive green – environmentally aware but makes no practical effort.
- Pragmatic green – aware of environmental and business trade-offs.
- Dark green – working towards a sustainable society.

Consider also that, as environmental management in business is a relatively new concept, existing middle and senior managers have been trained in traditional command and control management methods with maximised profits as the main driver. These people, like most others, have a tendency to stick with the tried and tested management theories that they have grown up with, although many are beginning to consider other management methods. Younger managers, especially those that have recently left training, are more aware of business activities and how they have an impact upon the environment. In general, younger managers will accept environmental management systems more readily as an additional tool for modern business management.

In this context, you also need to keep in mind that you will fall into one of the above five classifications yourself. Be honest with yourself and try to see which one you are. As you progress with the installation of the environmental management system, you and others within the company will change classifications, hopefully for the better.

4   The environmental audit

An environmental audit is a process of extracting information about a company that, when analysed, will provide a realistic assessment of how the company affects the environment and will also provide a set of environmental objectives and targets to reduce the effects. The establishment of the objectives and targets will form the basis of a corporate environmental plan.

Two audits will be encountered during the course of your company's quest for an environmental management system. The first audit – the environmental audit, the one that will be explained in this chapter – is the 'getting to know what a company actually does and how it affects the environment' audit. The second audit is a check to see whether an environmental management system works according to its procedures, and that objectives and targets are being achieved. The second audit, the 'system' audit, is explained further in Chapter 7.

Depending on your preference, or your needs, these audits can be as simple or as complicated as you, the implementer, are comfortable with, bearing in mind the minimum ISO 14001 requirements. In an attempt to satisfy the majority of readers, this chapter includes a comprehensive range of factors that can be assessed as part of an audit. Pick and choose what you consider to be appropriate. The lists provided are to save time in constructing your own. If appropriate, use your existing quality or marketing audits and integrate them with your environmental requirements.

An audit is the construction of the management equivalent of a spider's web (Figure 4.1). At the centre of the web is the environmental audit, which assesses – through its many information strands – where your company is now and the internal and external forces and factors that may or may not affect it in the future.

An environmental management audit is often lengthy; the implications are summarised as part of an aspects and impacts analysis. Therefore, an environmental management audit is not usually included as part of an environmental management plan, presentation or report. If it is required in the report, it should be included as an appendix at the end of the plan.

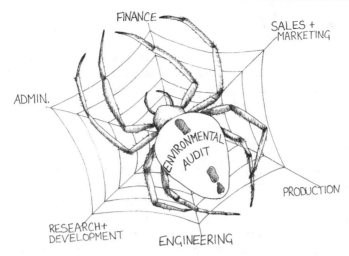

FINANCE

SALES + MARKETING

ADMIN.

ENVIRONMENTAL AUDIT

PRODUCTION

RESEARCH + DEVELOPMENT

ENGINEERING

Figure 4.1 An audit is the construction of the management equivalent of a spider's web.

Any environmental management audit includes an analysis of external as well as internal factors. External factors may arise from the environment, customers and/or competition. They relate to circumstances that your company cannot control directly and include broad factors such as the economic, customer, competitor and operational policy. The acronym ECCO has been used to aid external audit considerations. You must also include a breakdown of internal factors and analyse the resources and skills of your company, including any deficiencies or weaknesses.

This chapter is divided into seven broad sections. When you have completed them, you will have the foundation of your eventual environmental management system:

- the internal audit;
- the external audit;
- list of aspects and impacts;
- list of applicable legislation;
- environmental objectives and targets;
- environmental policy statement;
- corporate environmental plan.

Each section requires the identification of forces or factors that will have a major impact on the development of future environmental management plans or strategies, and the discussion of the future trends or changes in these factors. The impact that these changes or trends will have on a business or department will need to be outlined.

Hint

The internal audit should be carried out first. The benefit of this tactic is that determining what your company does and the impact its business activities have on the environment will allow you to identify quickly and clearly most of the external factors that affect, or may potentially affect, a company.

The need for auditing

At this point, it is worth a brief discussion as to why an audit is needed. Most people in business will have their own impression of what constitutes an audit. The audit process for an environmental management system has two parts. First, the pre-environmental management system audit provides knowledge of what a business's activities are and what impact they have on the environment. Some of these activities will have a positive or a negative impact upon the environment. The information gained from this audit will provide evidence for the chosen environmental management objectives and targets.

The second audit undertaken is designed to ensure that the environmental management system and its associated procedures are in place and that they actually work. For a representation of the implementation phases of an environmental management system and a demonstration of the auditing cycle, see Figure 4.5.

Bear in mind that, after the implementation phase, sufficient time must be allowed for the environmental management system and its procedures to function before the final audit is undertaken. For example, if you are pursuing ISO 14001, you must run your environmental management system for a 3-month period at least before final auditing.

Internal environmental audit

The first step is to list the key factors that the internal analysis might encompass and then conduct an internal appraisal. The results of this internal analysis will be used to form the basis of an aspects and impacts analysis that will be explained, in detail, later in this chapter.

Factors in an internal appraisal

As implementer of the environmental management system (EMS), you will need to identify the most relevant areas and activities to appraise in the company. Use your discretion and judgement in selecting areas to appraise, as the most relevant internal factors will vary among industries, companies and individual departments. A useful starting point is to list the major areas of company and departmental activities and the resources that are critical to competitive success. Essentially, these are activities and resources that result in outputs that are valued by existing customers, and it may be useful to ask their views as an aid in the identification process.

A check-list of some of the factors and areas which the internal part of the environmental management audit may encompass includes the following.

Financial

- Liquidity.
- Profit margins.

It almost goes without saying that, although the exercise of implementing an environmental management system will cost money, it is not going to involve major capital expenditure. It is, however, prudent to check on the extent of the cost of implementation because it will vary from company to company and from industry sector to industry sector. The cost of implementing an environmental management system in relative terms is not particularly high. If you want to calculate the cost to the last penny, the budget framework in Chapter 7 will provide some direction as to what to be aware of.

Personnel

- Managerial experience and expertise.
- Levels of training and education.
- Motivation and attitudes.
- Workforce skills.

Taking the time to identify the gaps in training, motivation, available skills, etc. and establishing development programmes will aid swifter implementation. There will probably already be a quality, health and safety or similar training structure to which environmental training can be effectively added, saving on company resources and minimising operational disruption.

Research and development (R & D) and design

- Budgets.
- Innovative success.
- Design expertise.
- Technological expertise.

Using a company's R & D and design capability, consider the likely results of making environmental changes to existing products and services. Will environmental changes allow the opportunity of offering better products to the market, or will a new, or modified, process offer additional savings or an additional by-product? Many by-products have emerged from companies looking at alternative uses for their product or process waste.

Engineering and production

- Production planning and control systems.
- Degree of automation.
- Quality control procedures.
- Age profile of plant.
- Flexibility.
- Unit costs of production.
- Supply and procurement.

Focusing on the engineering and production activities of a company may identify current and future challenges or present new opportunities. Consider, for example, some of the following areas for investigation.

1   Technology – materials, components, and machines.
2   Techniques – methods, systems.
3   Design, promotion, presentation.
4   Changes in legislation.

Closer inspection of existing methods of operation and the identification of alternative uses or changes to the production processes or engineering methods can bring significant benefits to material usage and energy efficiency. The case study in Box 4.1 illustrates this point.

Environmental management

- Environmental research and information systems.
- Environmental planning systems.
- Staff attitudes.
- Company image.

While auditing the company, make a check for the existence of environmental management skills and experience, taking note of any environmental projects that the company or individual employees have been involved with. Identifying those employees with relevant environmental skills will help when recruiting a working group.

Techniques of environmental assessment

Whether you are part of an SME or a larger company, there are many techniques open to you for environmental assessment. Employee

Box 4.1 Improving material usage and energy efficiency

Example – re-engineering existing power plant

There are many engineering and production issues to be addressed when environmental considerations are taken into account. It is more than 'out with the old technology and in with the new'. The current power plant debate is a case in point.

Power plants in the energy generation industry, particularly steam power plants, were designed to be operational for 20–30 years. Some could probably continue for a further 20–30 years with a few modern adaptations. The argument is that there are a number of options for an ageing power plant if engineering and production design and methods are revisited, combined or reworked in the context of today's technological advancement.

By examining different engineering techniques, several options emerge, from which selections can be made that will improve the power plant and reduce its impact upon the environment. These engineering options would include: repair, retrofit, rehabilitate, repower and co-generate.

Repair: recovery or improvement of original operational condition by applying modern, compatible technology.

Retrofit: the retrofitting of components using state-of-the-art technology can improve the overall efficiency of the unit.

Rehabilitation: having the capability to install new automated systems will extend the useful lifetime of the plant and reduce emissions.

Repowering: the introduction of other power units into a combined cycle system will improve performance, extend the system's lifetime and reduce emission levels.

Co-generation: modernising existing power plants to provide multiple energy-generating options. An example would be one power unit providing power to an industrial process and simultaneously producing electrical power for local community use.

---

interviews, focus groups (discussion groups) and questionnaires are three of the more common and easier methods of gathering environmental information. A combination of techniques can, of course, be used.

Employee interviews

Depending on the size of the company, all employees can be interviewed or just a cross-section of them. If it is to be a cross-section, make certain that every employee level and department is represented. This will ensure a range of views and a range of ideas for environmental improvement. The interviews can be either structured, i.e. you seek answers to specific operational questions, or unstructured, whereby just a few open-ended questions are asked and employees express their views across a broad spectrum of subjects relating to the question.

Focus groups

Formed from a representative cross-section of the company, a focus group discusses the subject of environmental issues within the company. You as the environmental manager will take notes of agreement and areas of discussion that come from the group. The size of the group may vary from small to a large company get-together.

Employee survey

A simple one-page survey or questionnaire with a few questions about operations and environmental impact can be very effective and very easy to administer. A simpler option, but one that can have just as much benefit, is a memorandum asking everyone in the company what environmental improvements they would like to see introduced. The memorandum approach is less structured and will elicit all kinds of responses from employees: some environmental, some health and safety, and even the odd grievance. While all the information received is useful towards understanding a company, the non-relevant responses, such as suggestions to increase available parking spaces, will need to be sifted out. The useable environmental suggestions will form the basis of the environmental objectives and targets.

The use of surveys in the environmental management audit can ensure a systematic approach and at the same time ensure that individual process or product managers are not, for example, avoiding aspects of environmental responsibility that may show them in a poor light.

External environmental audit

Identify the key external environmental factors pertinent to the business or department. You may consider that, knowing the company as you do, you can list the external factors that affect the business by heart. If this is the case, write them down and compare them with the lists given below.

By definition, 'the relevant external environment' encompasses all of the forces and factors outside a business that have some impact on it but that cannot influence or control it. The first problem is to identify the key factors pertinent to a particular department or unit and, where appropriate, a business.

In broad terms, it is possible to identify distinct groups of environmental factors, which, to a greater or lesser degree, potentially affect all industries and organisations. The acronym ECCO has been used to classify groups of external factors:

- *e*nvironmental pressure groups;
- *c*onsumers;

- competitors;
- operational policy (including legal and regulatory factors).

Below are some of the elements that need to be considered under each broad category.

Environmental pressure groups

- Friends of the Earth;
- Greenpeace;
- Council for the Protection of Rural England;
- The Royal Society for Nature Conservation;
- Local pressure groups can also be very important.

Although these groups may apply pressure in different ways, they are all very strong in the lobbying arena. They are not afraid to take on the largest of companies and they are adept at coming out on top in media battles. Witness the media exchange between Shell and Greenpeace on the scrapping of the Brent Spar. It was arguably a disastrous public relations exercise for Shell, and the proposed dumping of the Brent Spar at sea was abandoned, in keeping with public demand, in favour of the Greenpeace-initiated land-based disassembly. These days, multinational companies look to creating working partnerships with environmental groups and seek environmental input from them before actions are taken.

Consumers

- Economic growth.
- Income levels.
- Interest rates.
- Exchange rates.
- Balance of payment levels.
- Employment.
- Credit policies.
- Income distribution.
- Savings and debt.
- Taxation.

As consumer awareness grows of the environmental damage that can be caused by industry, people are being drawn, mostly by an ethical view, to purchase those products and services that are environmentally friendly, or those that do less environmental damage than others. Banks, financial institutions and various other company stakeholders are becoming more environmentally aware about where they invest money.

One could argue that environmental issues, as well as social welfare issues, come to the fore during periods of economic prosperity, and then return to a lower priority during periods of economic recession. This statement may have been valid in the early days of environmental management. There is now clear evidence to suggest that there is an acceptance by most of us that being more responsible for the earth's limited resources and how they are used will ensure that resources are sustained and continue to meet the needs of future generations.

Competitors

- The changing age structure of the population.
- Trends in family size.
- Changes in amount and nature of leisure time.
- Changes in attitude towards health and lifestyles.
- Improved education.
- Changes in attitudes towards family roles.
- Changing work patterns.
- Equal opportunities.
- Culture.

As competition for products and services grows, so competitors strive to be uppermost in consumers' minds. The identification of shopping patterns and consumer behaviour is an important tool for winning customers. The use of sophisticated software to monitor consumers' buying habits is an attempt to maintain or create an advantage over a competitor. Identifying and anticipating changing lifestyles, such as the increased expenditure on organic food by environmentally conscious consumers, adds value to producers' products and services if they are in tune with attitude changes.

It is sometimes a great temptation for companies, considering the immediate benefits of having a positive environmental image, to cut short the process and simply to change labels on their products and call them environmentally friendly. The discovery of such a practice can not only result in lost sales and market share but also incur greater expenditure (from recalling and relabelling products and possibly fines for mis-selling) than would a change to the product or the manufacturing process.

Operational policy

- Automation.
- New methods of travel.
- New materials.
- Improved communication.

The Internet, mobile phones and e-commerce are some examples of how a company may improve its business communications. Automation of processes may lead to greater energy efficiency and waste reduction. New environmental legislation may result in harmful products such as chlorofluorocarbons (CFCs) being banned from the production process. These are just some of the operational policy changes that need to be monitored and assessed for environmental impact. Some of these changes will have a positive impact upon the environment and some will have a negative impact. A reduction in car usage to and from work will have a positive environmental impact and will save people money, particularly at today's petrol prices!

It is not possible to list each and every external environmental element that may result in potential opportunities or threats in this part of the environmental audit. You must, however, determine the key environmental forces and factors that need to be assessed, and realise that these may vary between departments. Be pragmatic when undertaking this exercise. The smaller the company or department, the less of an impact these external factors will have. It is, however, important to give some time to this section.

Sometimes, the most relevant external environmental forces are not immediately obvious as new forces may develop or emerge, and existing factors can change very rapidly. It is therefore prudent to keep a broad perspective on what might constitute significant environmental forces and factors and review environmental plans on a regular basis.

The trends and changes in the categories of external environmental factors that you have outlined in the previous section can now be listed. Because environmental plans are developed to help compete successfully in the future, it is important to try to forecast both the magnitude and direction of trends and changes in those external environmental factors that have been identified as most significant. Use as many sources of information as possible to help forecast possible changes, including secondary data such as information found in government or industry association statistics and trade directories.

The time-scale that such forecasts will encompass is important and will vary greatly depending on the sensitivity of the industry in which you operate. As a rule of thumb, the time horizon for external environmental forecasting should be approximately twice as long as the duration of the environmental plan.

It is often said that the essence of effective environmental management involves achieving a strategic fit between organisational activities and the environmental threats that exist in the environment in which you operate. In your environmental management audit, you should have identified key external factors, and trends and changes in these factors, as well as the performance and resources of your organisation with respect to internal factors. The mechanism for moving from the information provided in your environmental management audit to using this information in developing an environmental management plan is provided by the aspects and impacts analysis. Essentially, such an analysis is used to develop a plan that builds on identified strengths and avoids or obviates environmental risk.

Aspects and impacts analysis

Before entering into this analysis, it is worth identifying what an 'aspect' is and what an 'impact' is; the following definitions may help. For the sake of consistency and standardisation, the definitions offered here are those laid down in the ISO 14000 standard. If you are pursuing ISO 14001, you will use those definitions and guidelines stipulated in the ISO 14000 series.

ISO 14001 guidelines describe an environmental aspect as:

> ... an element of an organisation's activities, products or services that can have a beneficial or adverse impact on the environment ...

and an environmental impact as:

> ... the change that takes place in the environment as a result of the aspect ... .

The register (Figure 4.2) lists the details of any impacts under the following headings:

Item:	a serial number that is used for recording and sorting purposes.
Environmental aspect:	brief details of the company's activity, product or service that can have an impact on the environment.
Environmental impact:	brief description of the interaction or change.
Caused by:	details of the operation, activity or process that caused the environmental interaction or change.
Effect:	details of whether the environmental interaction or change has an adverse or beneficial environmental effect.

Item	Activity or process (aspect)	Environmental effect (impact)	Caused by	Effect

Figure 4.2 Register of aspects and impacts.

Aspects and impacts examples and description

When it has been established what the company does, all the activities and processes should be listed and the list divided into three key areas. First, those activities and processes that affect the environment during normal operating procedures should be identified. Second, all those activities and processes that affect the environment during abnormal procedures should be identified; and, third, the same for incidents, accidents and other emergencies.

• Normal operations – daily operational activities or processes that are currently being carried out.
• Abnormal operations – periodic routines that may occur in addition to the daily process and activities. These may include preventive maintenance, plant upgrades, shutdowns or silent periods.
• Emergencies – at best, when the unexpected happens, this may be a small, localised spill or breakdown. Or, at worst, this may be when the hounds of hell are unleashed as your 3000-litre heating oil storage unit ruptures and its contents disappear into the ground. In the context of environmental disasters such as Chernobyl and the *Exxon Valdez*, these lesser environmental emergencies would pale into insignificance – unless of course they happen to your company.

Having prepared a list of the company's activities and processes (aspects) and having identified the environmental impacts (both positive and negative) of these, it is useful to introduce a measure to determine their significance (Figure 4.3). Those aspects that have a significant impact upon the environment will need to be identified and prioritised.

Aspects and impacts analysis matrix

For ease of explanation, a five-point rating scale is used to classify each aspect's environmental impact as major, high, moderate, limited or minimal. The measurement criteria for each impact classification are shown in Table 4.1.

A simple formula for an aspect's impact rating (Table 4.2) would be:

Impact rating = aspect weighting × impact weighting

To determine the impact rating of your process or activity, judgement should be used initially to assess the level of resource used or the level of emissions to the environment. Calculating the levels of 'low', 'occasional', 'moderate', etc. will be an assessment based on existing Environment Agency pollution acceptance levels, your experience and the company size. In determining the level of resources used, bear in mind the types of

Item	Activity/product/service	Environmental aspect	Environmental impact	Positive/negative
01	Use of tinning powder	Potential for spillage	Possible atmospheric contamination	Negative
02	Use of tinning powder	Possibility of accidental spillage	Possible soil and water contamination	Negative
03	Use of copper plate	Direct flow to drains	Water contamination	Negative
04	Use of Freon solution	Direct flow to drains	Environmentally approved washing liquid	Positive
05	Generation of waste paper	Recycle waste paper	Conservation of natural resources	Positive
06	Preparation of wire and cable assemblies	Reuse of wire 'tails' and offcuts	Reduce product costs	Positive
07	Servicing of company cars	Potential for oil spillage	Possible soil and water contamination	Negative
08	Servicing of company cars	Exhaust emissions	Reduction of atmospheric contamination	Negative
09	Replenishment of oil-fired space heaters	Possibility of accidental spillage of fuel oil	Possible soil and water contamination	Negative
10	Packaging	Reuse of packaging material	Conservation of natural resources	Positive
11	Engraving	Plastic and metallic airborne fibres and dust	Possible atmospheric contamination	Negative
12	Engraving	Use of air compressor	Noise pollution	Negative
13	Soldering PCBs	Use of lead-based solder paste	Possible atmospheric contamination	Negative
14	Assembling PCBs	Use of air compressor	Noise pollution	Negative
15	Heating oil storage	Use of fuel tank bund	Possible soil and water contamination	Negative
16	Inherited waste oil tank	Waste oil tank leakage	Soil and possible water contamination	Negative
17	Inherited oil-contaminated soil	Soil contamination	Soil and possible water contamination	Negative

Figure 4.3 Identifying positive and negative aspects and impacts.

Table 4.1 Impact classification

Rating	Impact classification	Impact criteria
1	Minimal	No noticeable environmental effect Effective control system already in place Well within discharge consent levels
2	Low	Low environmental effect Substantial control measures in place to limit impact
3	Moderate	Known effect upon the environment Limited measures in place to handle impact Infrequent monitoring undertaken
4	High	Noticeable effect upon the environment Minimal measure in place to handle impact Haphazard monitoring undertaken
5	Major	Highly noticeable effect upon the environment No measures in place to limit impact No monitoring undertaken

Table 4.2 Aspect classification

Rating	Aspect classification	Aspect criteria
1	Minimal	No emissions and no use of resources No hazardous material usage
2	Low	Low emissions and low usage of resources Occasional use of hazardous materials
3	Moderate	Moderate emissions and use of resources Moderate use of hazardous materials
4	High	High emissions and high use of resources High use of hazardous materials
5	Major	Major emissions and major use of resources Major use of hazardous material

materials employed. The use of solder paste, particularly lead-based solder paste used to secure components in printed circuit board (PCB) assembly, would typically carry a high aspect weighting. Be aware also that emissions can be to land, air and water.

To complete your toolbox for the matrix application, you will need to assign some degree of probability to the likelihood of the aspects and impacts occurring. Classification of an aspect can be according to Table 4.3. The probability of occurrence can be measured from a score of 1 (less than 20 per cent probability) to a score of 5 (an 81–100 per cent probability

Table 4.3 Probability of occurrence

Probability of occurrence	Probability (%)
5	81–100
4	61–80
3	41–60
2	21–40
1	0–20

Aspect	Probability of occurrence	Potential impact score Very good								Very bad			Total
		5	4	3	2	1	0	−1	−2	−3	−4	−5	
Chemical spillage	5											X	−25
Exhaust fumes	5								X				−10
Product packaging	5		X										20

Figure 4.4 Aspect and impact probability matrix.

of occurrence). The calculated probability of occurrence scores are transferred to the 'potential aspect/impact/probability matrix (Figure 4.4).

By multiplying together the probability and impact scores, it is possible to determine whether the aspect has significant environmental impact. If the aspect score is less than 9, the aspect can be considered as not significant. If the score is 9 or greater, the aspect can be considered significant. Note that each process or activity may have a number of associated aspects and impacts.

Understanding the matrix.

The aspects that have been identified in your survey should be listed in column 1 of the matrix shown in Figure 4.4. In the next column, you should enter a number, 1–5, to represent the percentage probability of the aspect occurring. The following columns, ranging from 5 to –5, represent the scoring system for the aspect having a positive or negative impact upon the environment. The examples given in the matrix demonstrate, first, a negative impact – a chemical spillage. A company that uses any chemicals in its operating process is likely to view any spillage as being serious, therefore a –5 score would be applied to this aspect of operations.

In contrast, the reuse of packaging materials received from incoming goods can be viewed as a positive environmental impact because the same packaging materials can be used for packaging outgoing products. This would realise a score of 5 to signal a positive environmental impact.

By multiplying the scores together, the figure in the last column will determine each aspect's impact upon the environment. The larger the negative figure, the larger the environmental impact. The larger the positive figure, the less of an impact upon the environment.

Keep in mind that an aspect can have both positive and negative impacts upon the environment. For instance, using the packaging example, if you received a significant amount of paper packaging from incoming goods and did not reuse it, the recycling of this packaging waste would be a positive aspect but the money you spend to have the packaging recycled may be a negative aspect.

Whether a company is a freighting or service company, or a large or small employer, will determine the significance of particular aspects. The aspect of exhaust fumes, for example, is likely to appear on most company registers, and the decision as to whether it is a significant aspect is your responsibility, as implementer, and your assessor's. If you have 'missed' an aspect, or wrongly assessed its impact, your assessor will ask you to reconsider what an aspect is and also its likely impact upon the environment. Rather than tell you directly, the assessor will go back through the aspect assessment process with you and question whether you have considered every eventuality.

Departmental analysis matrix

Without organising larger businesses into separate departments, it is virtually impossible to develop meaningful business definitions or meaningful and effective environmental management plans. It is extremely important to stress that separate parts of an environmental management plan must be developed for every department in the organisation. In other words, the development of subsequent parts of an environmental management plan must be undertaken at the level of each department.

The departmental concept enables each part of a business to have its own business definition. For every department, a business definition should specify the following elements:

- the environmental objectives for each department;
- the resources available to meet the stated objectives;
- the technology to be utilised in product or process changes.

A good way of thinking about the implications and magnitude of your opportunities and threats is by using the chart shown in Figure 4.5. Specific values at the foot of the chart will correspond with the total scores for

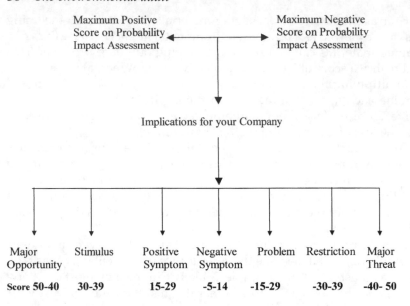

Figure 4.5 Impact opportunity and threat analysis.

each factor on your potential impact/probability matrix and indicate the potential implications of each factor on your company or department.

It is also useful to conduct an analysis that helps you to understand the strengths and weaknesses of your company's competitors. Use percentages to identify and weight the relative importance of the key success factors in your company's industry. When all weights are added together, the total should add up to 100. For each success factor, you must then rate your company and your two or three major competitors on a scale from 1 (a low or poor competitiveness rating) to 10 (a high or excellent competitive rating). Adjusted competitiveness ratings are calculated by multiplying the weighting factor by the individual company's ratings. Total company competitiveness ratings are calculated by adding the adjusted ratings for each success factor together for each company. (To determine competitor ratings, use the departmental analysis shown in Figure 4.6 and change the 'department' classification to 'competitors'.)

Figure 4.6 illustrates a simple departmental analysis table for the hypothetical aspect/impact analysis matrix described previously. It contains only three environmental factors; in reality, it is likely that there would be many more. An examination of Figure 4.6 might show that your company's key environmental factor lies in the area of waste generation. For example, if department B is – according to company waste records – the largest waste generator in the company, you may consider that this area should attract a larger portion of available resources.

Give rating from 1 (lowest or poor) to 10 (highest or excellent)						
Key environmental factors	Weighting factor	Your business		Department A		Department B	
		Rating	Adjusted rating	Rating	Adjusted rating	Rating	Adjusted rating
Waste generation	50	8	*400*	3	*150*	7	*350*
Energy usage							
Air emissions							
Total							

Figure 4.6 Departmental aspects and impacts analysis.

Summary

Having identified the environmental aspects of the company's operations and calculated their impact upon the environment, it is worth reflecting on how the information can be put to good use. The information that will have been gained is detailed briefly below:

- Those areas of company operations that may harm the environment have been identified (identifying the aspect).
- The different aspects of the company's operations have been classified into levels of environmental impact, i.e. low, medium or high.
- A scale of 1–5 has been used to determine each impact's probability of occurrence.
- A matrix has been created to calculate the impact scores for each aspect.
- Those aspects above the level of a 'mid' score, or a cut-off score, can be selected and used to create a register of aspects and impacts.

If you have aspects and impacts to be determined on a departmental basis, repeat the process above for each department and compare departmental scores (see Figure 4.6).

Having created a register of aspects and impacts, attention should now be paid to the creation of the environmental legislation register.

Register of environmental legislation

It is essential that, as implementer of the EMS, you are fully familiar, and compliant, with the regulations that apply to the company. A fundamental part of the planning and implementation process is the creation and on-going maintenance of a register of legislation and regulations. The register is explained below, and for ease of reference your register should be kept in your environmental documentation folder (see Chapter 10).

You should maintain a register of those regulations that apply to the environmental aspects of company activities. The register lists the regulations under the following headings:

Item:	A serial number, specified by you, that is used for recording and sorting purposes.
Regulation:	Title and brief details of the regulation.
Issued by:	Details of the authority that has issued the regulation.
Dated:	The regulation issue date.
Applicability:	Details of your operations, activities or processes subject to the regulation.

Your EMS requires that a register be kept of the various items of legislation and regulation that are applicable to your company. A useful list of environmental legislation that could possibly apply to your company is detailed in Appendix 2. The skill to be mastered is to determine which pieces of legislation affect your company's operations. There are two key acts that will apply to all companies, the Environmental Protection Act 1990 and the Environmental Act 1995. These two acts will be in every company's register.

Establish objectives and targets

An analysis of the responses given in the feedback questionnaire should provide sufficient information to allow the determination of the company's objectives and targets.

The setting of environmental objectives is the main way in which a company's environmental performance can be improved. The objectives set do not need to be quantifiable, but they do need to be realistic and achievable. The targets are the quantifiable part, but it may not always be easy to calculate a target, particularly if this is new to you. Be conservative in estimating savings. It will be difficult to determine exactly how much energy has been saved or how much material has been reused until the environmental management system has been operating for a full year. So think of an annual percentage saving on the company's targets of between 5 per cent and 10 per cent. This percentage will be based on knowledge of the company and adjusted once an adequate measurement period has passed – possibly a year.

Table of objectives and targets

The setting of environmental objectives is the main way that a company's environmental performance can be improved. The objectives of the EMS should encourage all employees to operate as part of an environmentally caring company, and this should be demonstrated by the employment of an effective environmental management system.

The primary objective is to operate and maintain the company in a manner consistent with the best environmental practices, taking account of responsibilities to customers, staff, suppliers and the community at large. The environmental plan (EP) is designed to be open-ended so that new, or revised, objectives can be added to, or changed within, the programme. In the early stages, this should be on a fairly regular basis.

Targets

The environmental management plan is designed to be a dynamic and open-ended list of objectives and goals for a company to achieve in accordance with its stated objectives and targets.

The EP should be regularly reviewed and updated by the environmental manager in consultation with the environmental working group. The managing director should also carry out a periodic review of the EP. Periodic progress reports should be circulated within the company.

The EP takes the form of a list of targets that is designed to be as simple as possible while showing complete details of goals, methods and responsibilities. The active EP target details, again for ease of reference, should be contained in the environmental documentation folder (EDF).

EP targets list

The EP target details are contained in the targets list under the following headings:

Item:	A unique item number is assigned to each objective; this acts as a reference in reports or reviews.
Objective:	The specific objective is described and detailed. Some objectives may require more information; if so, this will be provided separately and a copy placed in the relevant objective's folder.
Target:	For example, reduce energy consumption by 10 per cent.
Goal:	In this context, a goal is the projected target date for completion of the objective. There needs to be flexibility built into these dates; consequently, the dates may be altered following a progress review – as may the objectives and targets.
Achieved:	This is the actual objective achievement date.
Responsibilities:	Each objective will have a person or persons designated as being responsible.
Comments:	This area is reserved for any pertinent comments.

Environmental policy statement

Within the scope of a company's activities, products and services, senior management should support a signed environmental policy statement that states the company's commitment to protection of the environment, prevention of pollution and continuous improvement of company environmental performance. For example, these aims could be to:

- reduce consumption of energy and water;
- reduce generation of waste;
- reduce emissions.

Senior management should also verify what the company will do to meet the stated aims, and may identify some of the following examples:

- Provide adequate resources and personnel to maintain the EMS.
- Ensure that employees are educated and trained to understand their responsibilities in respect of the environmental policy.
- Ensure that the requirements of environmental legislation are met and, where possible, exceeded.
- Provide a framework for setting environmental objectives and targets.
- Integrate environmental considerations into the design of products and services in order to avoid or minimise environmental impacts.
- Monitor environmental performance continuously.
- Review and audit the effectiveness of the company environmental policy.
- Where it is feasible, work with suppliers, contractors and sub-contractors to improve their environmental performance.

This statement should be signed and dated by the managing director and another director or senior manager to demonstrate senior management support for the initiative.

Corporate environmental plan

The corporate environmental plan (Figure 4.7) is the culmination of the auditing process. Having identified, listed, analysed and prioritised the operational aspects of the company, the environmental impacts, emergent targets and objectives will form the basis of the corporate environmental plan. A corporate environmental plan should be designed as a dynamic and open-ended list of objectives and targets. Some of the initial targets and objectives that are set in the early stages of introducing an environmental management system will be realised when accreditation of ISO 14001 has been achieved. One of the first objectives listed may be the achievement of an environmental management standard.

Item	Objective	Target	Method	Responsibility	Goal	Status	Date
01	Tracking and recording recyclable paper and cardboard	Following the establishment of a base figure by Dec 02, a minimum of 5% reduction in recyclable paper and cardboard will be set	Gradually increasing the amount of paper and cardboard that is recycled	Procurement manager	5% reduction	Ongoing	Dec 01
02	Reduction in office and workshop paper and cardboard waste output	Following the establishment of a base figure by Dec 02, a minimum of 5% reduction in paper and cardboard waste will be set	Use of memorandums and newsletters to advise personnel on the need to reduce paper and cardboard usage				
03	Reduction in office and workshop plastic waste output	Following the establishment of a base figure by Dec 02, a minimum of 5% reduction in office and workshop plastic waste disposal will be set	Use of memorandums and newsletters to advise personnel on the need to reduce plastic				
04	Reduction in workshop wooden and metallic waste output	To establish a means of quantifying a reduction in workshop wood and metal waste	Use of memorandums and newsletters to advise personnel on the need to reduce wood and metal waste				
05	Reduction in energy consumption	Reduce energy consumption by a minimum of 5%. To include electricity and heating fuel. Baseline to be determined by Dec 02	Personnel encouraged to make savings where possible. Lower thermostat settings and more efficient heaters				
06	Reduction of hazardous materials and substances	To establish a means of quantifying a reduction in hazardous materials and substances output	Use of memorandums and newsletters to advise personnel on the need to reduce the use of hazardous materials and substances in products and processes				
07	Reduction in water consumption	To establish a means of quantifying a reduction in water consumption	Use of memorandums and newsletters to advise personnel on the need to reduce water consumption				

Figure 4.7 Corporate environmental plan (CEP).

In short, some objectives and targets will have completion deadlines, and when achieved these should be deleted from the corporate plan and replaced with new objectives and targets. As the management review is carried out on an annual basis, operational activities may have changed during this time and this will necessitate the introduction of new targets and objectives. The process of reviewing and restating the environmental objectives and targets offers the opportunity for continuous operational and environmental improvements to be made.

5 Environmental plan structure

The intricacies of the environmental plan are outlined in this chapter. First, an explanation is provided of the key functions of elements within the plan that will ensure that an environmental system is installed effectively and operates efficiently. This is followed by a brief overview of some of the key points to be considered when a plan is presented as a working document to senior management or other interested parties.

Implementing a new system within a company does need a good plan (Figure 5.1), but it is also useful to have an understanding of how companies and people work. This chapter also offers insight, therefore, into how companies learn, what barriers can emerge to new systems and how the political intricacies of individuals can be identified; hopefully, with this type of identification comes an understanding that will aid the implementation of a new system.

Management commitment

To ensure the successful installation of an environmental management system, it is essential to engage full management commitment. The directive may come from above but, as implementer, you must ensure that commitment manifests itself in the form of people and budgets and, above all, you must ensure that one or more members of the senior management team are committed to the project. If possible, try to recruit one director onto the working group. Lack of support from above will make the implementation process difficult, if not impossible (Figure 5.2). There will be many political issues to overcome throughout the implementation process and you will make faster headway if a director handles the political negotiations.

The environmental audit

The environmental audit requires the current level of environmental risk to be analysed and shows where a company stands in relation to the risk. The audit is a prelude to determining a company's future objectives (where

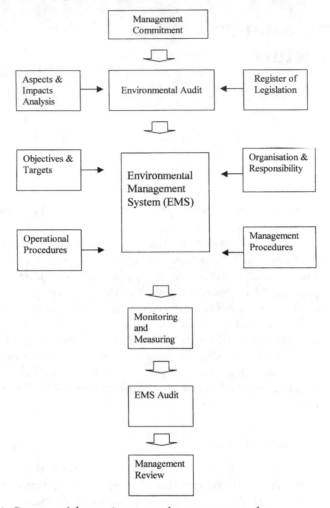

Figure 5.1 Content of the environmental management plan.

it wants to be) and the procedures for achieving these objectives (how it is going to get there).

The environmental audit will provide an insight into how a company operates. If an auditing process is already in place, it should be applied in the environmental context. An audit is a natural management tool; they are carried out every day – management and employees take stock of what is currently taking place within a company and see how events can be improved or applied in different ways to enhance company performance. Exactly the same principles can be applied to the environmental audit. The four main areas of the environmental audit are:

Figure 5.2 Lack of support from senior management will make the implementation process difficult.

- internal audit;
- external audit;
- aspects and impacts analysis;
- register of legislation.

Internal audit

An internal environmental audit needs to be conducted to assess what products and services the company currently produces or provides. If it is clear what the company does, this knowledge should be taken one step further and, as implementer of the EMS, you should become aware of the impacts that the company's products, processes and activities have on the environment. Once this is done, a list of the operating aspects of the company that have an impact upon the environment will have been created. The next step is to determine the level of impact that each aspect has and how significant it is to a company's operations.

External audit

In this phase, the environmental issues that currently, or have the future potential to, affect a business must be considered. Chapter 4 uses the ECCO acronym to direct attention to environmental issues that may arise as a result of changes in the economy, competitor or consumer activities and / or operating policy. The biggest issue for most companies at present is the growth in environmental legislation that has, or should have, an immediate impact on a company's operational policy and decision-making.

Aspects and impacts analysis

An important stage in improving a company's environmental management performance is the preparation of an aspects and impacts analysis arising from an assessment of the company's activities and processes. The identification of the environmental impacts arising from the environmental aspects will form the foundation of the targets and objectives for a corporate environmental plan.

The completed aspects and impacts analysis sheet will provide a record of the environmental effects of operating procedures, incidents, accidents and potential emergency situations that may arise. This will assist in the monitoring of their effects on the environment and the development of any necessary remedial actions and procedures. The aspects and impacts analysis provides a positive and negative record of all business activities that are significant environmentally. Chapter 9 will demonstrate how the significance of each aspect may be calculated.

The formulation of the aspects and impacts analysis sheet will be completed at an early stage of implementation; be aware that the analysis sheet should be updated as and when required. If new equipment is introduced into the company, or one of the processes is changed, the environmental impact should be assessed and its significance tested to determine whether it should also be included.

Register of legislation

As a result of the external audit, a list of current legislation can be compiled that has an actual or potential impact on the activities of a business. When searching through the list of legislation, all legislation that can affect a company should be taken into account, not just environmental laws. If you are carrying out the search and are unsure whether the legislation carries a risk to, or may have an impact on, the company, list it anyway and verify it later – better safe than sorry.

Having assessed both what a company does and its environmental implications, the company is now in a position to formulate its environmental management system. This begins with the creation of a list of objectives and targets, and the development of a set of management and operational procedures to ensure that the objectives and targets are achieved. The company structure and those individuals within it that will be the key persons to which the success of the system will be entrusted should also be assessed.

Objectives and targets

The achievement of, or the attempt to achieve, environmental objectives is the main way that a company's environmental performance can be improved. The objectives set do not need to be quantifiable, but they do need to be realistic, identifiable and achievable. The first group of objectives and targets that are likely to be set will reflect the introduction of the new system. As these short-term objectives are achieved, so new objectives will be set, thereby creating a dynamic group of objectives that takes a company through the process of continual improvement (Table 5.1).

Organisation and responsibility

Having identified what the company wants to achieve, consideration must now be given to how the company is structured and to the level of employee involvement, including the detailing of specific tasks and responsibilities. At this point, the introduction to the environmental management plan being written should be mentioned what the company does, where it is based and how long it has been in existence. The size of the company in terms of turnover and number of employees should also be mentioned.

The type and level of resources available for assessing how achievable a company's objectives and targets are should also be verified. At the same time, take the opportunity to detail the training and communication programmes designed to ensure staff awareness and education.

Operational procedures

The establishment of environmental operational procedures is the main way in which environmental objectives and targets can be delivered. They are the basis for ensuring that the requirements of the ISO 14001 standard are being adhered to. The control of the correct application of the operational procedures is maintained by environmental management procedures, and these should be written in conjunction with the person responsible for their operation.

Management procedures

The environmental management procedures control the operation of the EMS. They specify the way in which any and every particular activity should be undertaken. The management procedures come into operation when the operational procedures do not function correctly or are not being adhered to. If part of the EMS fails, the management procedures give guidance for reporting and correcting the failure.

Monitoring and measuring

An important part of establishing and maintaining the effectiveness of an environmental management plan is ensuring that the company's significant

Table 5.1 Objectives and targets format

Item	Objective	Target	Method	Status
01	Completion of draft EMS documentation	Completion of draft EMS documentation, including manual, procedures, logs and forms	Follow requirements of ISO 14001 and any available examples	Complete
02	Completion of final standard EMS documentation	Issue of final standard EMS documentation	Documentation to be finalised, reviewed, approved and issued following review of drafts	Ongoing
03	Accreditation assessment preparation	Successful completion of stage assessment	Follow requirements of ISO 14001	Ongoing
04	Premises clean-up. To include vehicles, tyres, containers, scrap items and oil contamination of soil and bund	Remove all potential contaminants and inherited contamination	Liaison with local council, clean-up agencies, scrap dealers and vehicle owners	Ongoing
05	Tracking and recording waste disposal, starting with recyclable paper and cardboard	Recycling and reusing as much paper and cardboard as possible	Identification of suitable haulier and use of waste collection log	Ongoing
06	Reduction in office and workshop paper and cardboard waste paper output	Noticeable reduction in waste paper and cardboard output	Publicity and placement of marked boxes throughout company; personnel encouraged to reuse where possible	Ongoing

07	Reduction in office and workshop plastic waste output	Noticeable reduction in waste plastic output throughout company; personnel encouraged to reuse where possible	Publicity and placement of marked boxes	Ongoing
08	Reduction in workshop wood and metal waste output	Reuse of material and minimising of waste output	Publicity and placement of marked boxes in company workshop; personnel encouraged to reuse where possible	Ongoing
09	Reduction in energy consumption	Reduce energy consumption by a minimum of 10%; to include electricity and heating fuel	Personnel encouraged to make savings where possible. Improve building insulation, replace heaters, install new ceiling	Ongoing
10	Reduction of hazardous material and substances	Noticeable reduction in use of hazardous material and substances	Personnel encouraged to identify and use environmentally friendly alternatives	Ongoing
11	Reduce the possibility of emission of contaminated air	A cleaner atmosphere	Replacement of inefficient heating system	Ongoing

Note
EMS, environmental management system.

environmental impacts are being controlled on a daily basis. The process of monitoring and measuring will make certain that information is received to ensure that this control and continuous improvement occur.

The issue and control of documentation is important to the effective monitoring and control of environmental procedures. Be vigilant, however. Do not reinvent the documentation wheel; if there are existing quality control documents, use them.

Environmental management system (EMS) audit

The primary outcome from an environmental system audit should be the capacity to assess whether the actual performance of the company, or departmental, environmental system conforms to the planned objectives and the requirements of the ISO 14001 standard. Any change, or recommendations for change, to the system will generally follow a system audit. Findings from the audit will be discussed and actioned at the management review.

Management review

If the company is to achieve continual environmental improvement, the environmental plan needs to be reviewed and examined regularly. The review should not be seen to be restrictive – it should encompass all aspects of the environmental management system. It should be viewed as a health check for the well-being of the system.

The management review will discuss every element of the EMS to ensure that each is working effectively. Future objectives and targets will be identified and discussed during the review. The commitment of senior management is likely to be most in evidence at the management review. All those senior managers present at the review should agree every decision made and the managing director should ratify the minutes of the management review.

Having completed the details of the plan according to the framework detailed earlier in this chapter, it is useful to prepare a budget to provide an indication of the cost of installing an environmental management system.

The environmental plan budget

The environmental management plan budget is crucial, and ultimately provides the foundation on which key strategic and funding decisions are based. Therefore, it is important that it is accurate and well presented.

The financial element of the plan should describe the environmental management budget and the ways in which it relates to the targets that were developed earlier in the environmental management plan. It should include a statement of the funds required, their intended purpose and the projected impact on the profitability of the business.

Company politics

Life would be easy if the preparation of a plan and the detailing of its budget costs were all that it took to begin implementing a new system – the complexities of company politics must also be taken into account.

It does not matter how much you try to ignore company politics, they are ever present. If you are implementing an environmental management system, or for that matter any system, you will need to be aware of the power and political interactions that occur between people in the workplace.

There are a number of political dimensions that need to be addressed when introducing an environmental management system. First, as implementer, your needs and motivations will need to be considered. Then, you will have to consider the needs of your working group, and, finally, how you and the group interact with the rest of the company.

So, to give yourself a political edge, try to identify the politicians within the company and how they operate. There are some obvious questions to ask yourself: How can the politicians within the company be identified? How do they differ in their methods of controlling and co-ordinating activities? How do they reward and motivate people? How do they influence and change their work environment? Finding the answers to these questions will also identify the type of politician that you are. The following descriptions of types of company politicians have been borrowed from Andrew Kakabadse's (1983) book *The Politics of Management*:

- the company baron;
- the visionary;
- the traditionalist;
- the team coach.

The company baron

Methods of controlling and co-ordinating

Company barons (Figure 5.3) are focused on two main objectives: improving the company and improving their own position within it. To achieve these ends, control is exercised through subordinates, who are made aware of the company baron's status and the subordinate's role in relation to that status. Control is maintained through superior–subordinate distance and ensuring that subordinates work within the, much emphasised, formal and informal role constraints.

Co-ordination is exercised through extensive interaction, or networking,

Figure 5.3 The company baron.

with other senior managers, work groups, committees, etc.

This approach to co-ordination is a positive move towards change, albeit controlled and steady within the company. If an individual were to be perceived as introducing change too quickly, or that change were perceived to be too radical, the change would simply be stopped because, in the company baron's opinion, the person or change concerned was upsetting the traditional practices that made the company successful.

Methods of reward and motivation

Rewards and motivation, as far as the company baron is concerned, are in the subordinate's own hands. If he or she follows the rules, does what is asked and is respectful, the company baron will argue for increased pay and promotion on the subordinate's behalf. That said, if the subordinate attempts to force either of these two issues, he or she risks losing the company baron's support.

Methods of influencing and changing

The company baron is not against change as long as the change is phased in small controllable stages. Trying to maintain the company's traditions and to control the rate of change requires a substantial balancing act involving a mixture of personal power and caution. The caution is used in an attempt to slow down the rate of change; the personal power is used to create informal networks to support the change policies of the company baron.

The visionary

Methods of controlling and co-ordinating

Although the visionary (Figure 5.4) and the company baron both work towards the long-term benefits of the company, the visionary works at a faster rate and welcomes rapid changes. The visionary likes to have other visionaries as part of the working group, which operates under a regime of direct control as opposed to co-ordination. Confrontation and conflict are tolerated as long as the job gets done. The visionary, unlike the company baron, is not a respecter of the system. If something is not going to plan, either someone or the system is at fault, and either or both will be changed.

Methods of reward and motivation

Money and status are bestowed on those individuals who can work with little guidance, can introduce change, can take calculated risks and are willing to accept increasing levels of responsibility. Despite the visionary's penchant for recruiting like-minded people to work within the group, little friendliness and loyalty will be displayed. There is one good reason for

Figure 5.4 The visionary.

this attitude: if the professionalism and policies of a team member do not fit with those of the visionary, the person concerned is ditched.

Methods of influencing and changing

The skills of the visionary are those of confrontation, direct control and personal charisma. The visionary rewards those who implement change efficiently and quickly. If there are restraints on the speed of change or on the change itself, the visionary will cast aside both friend and foe, and will start again to get the job done. The visionary has no time for anyone who identifies with the values of the past, or is negative or considered a block on progress.

The traditionalist

Methods of controlling and co-ordinating

Traditionalists (Figure 5.5) are people oriented. They seek group membership and look to control and co-ordinate people within the group. For the traditionalist, knowing who is doing what in the group and at what speed and whether it fits with what others are doing is important. Co-ordination and control are applied to any person that complies with the group norms.

To reprimand or bring the errant person into line, the traditionalist will provide closer supervision, more work and will reprimand if necessary; he or she will distance themselves until the individual begins again to show accepted group behaviour.

Methods of reward and motivation

Before the traditionalist begins to reward and motivate others, effective communication and comfortable interpersonal interaction has to be established. For this to be achieved, the communication has to be conducted formally through memorandums or prearranged meetings. Throughout all personal interaction, the traditionalist requires that the superior–subordinate distance is acknowledged and maintained. By adhering to these traditional values, the subordinate will be rewarded with paternalistic approval. For this approval, the traditionalist would expect thanks and a continued willingness to do further good work.

Methods of influencing and changing

In a world of ever-increasing change, the traditionalist seeks to maintain the status quo. In the face of new systems being introduced and new

Figure 5.5 The traditionalist.

working groups being formed, the traditionalist will resist such change strongly. At best, the traditionalist will slow the rate of change to a comfortable snail's pace. At worst, he or she may feel threatened and react in a negative fashion by boycotting meetings and rejecting new working methods.

The team coach

Methods for controlling and co-ordinating

Unlike the traditionalist, the team coach (Figure 5.6) is aware that it is impossible to maintain the status quo over the long term. The team coach realises that working with a variety of different types of people within the company is necessary to maintain growth and implement the resultant changes.

Control and co-ordination is exercised through the generation of rapport with all group and non-group members, and by administering understanding and sympathy. Meaningful working relationships are developed through the use of interpersonal skills. The emphasis is more on the individual and less on the work details.

Figure 5.6 The team coach.

Methods of rewarding and motivating

The team coach's main method of operation is to develop a greater openness with regard to the discussion of both personal and work problems. The approach is designed to create greater trust and rapport within the group and thereby a consensus decision-making process. Anyone who is involved with the group and who receives praise from the team coach for work well done will consider this reward in itself.

Methods of influencing and changing

The team coach adopts a flexible position for change compared with the traditionalist, providing it does not upset the working group significantly. Flexibility is also displayed between groups. If cutbacks are required and they apply to all working groups or departments within the company, then all is well with the team coach. On this basis, it may seem like the team coach is the ideal person to effect change. Decision making, however, can be a weakness, particularly when the going gets tough or the speed of change is quick and unfamiliar.

Most people would recognise that different types of people exist within every company and that, regardless of whether they like or dislike them, they have to interact with all types. It is unfortunate perhaps that different people do not stay within their own sphere of understanding. They endeavour to expand and influence others with their own values and judgements. In order to understand which of the company politicians are thriving, you must consider how the company learns and the pressures and forces that shape it.

The learning process

Having identified the politicians within the company, it would be useful to understand the ways in which we learn. The 'experiential learning model' presented by David Kolb (1984) represents a four-stage learning cycle (Figure 5.7).

According to Kolb, the learning process begins with concrete experience that is based upon observation and reflection. Our observations and reflections allow us to form abstract concepts of the world around us, as well as concepts and generalisations for future actions. The future actions are realised in the fourth stage of the learning process by testing our new theories and concepts in new situations. This provides us with further concrete experience and the process begins again.

Of course, we do not all learn in this circular fashion. Different situations throughout a working day will require a different element of the learning process to be brought into action. It is also conceivable that personal bias and discrimination can create barriers to the complete implementation of the learning proces. For instance, to enable a freer learning process for the individual it is necessary for the company to foster a learning culture. In other words, the learning objective should be explicit and encouraged, to improve both the individual and the company.

How do companies learn?

Most management consultants and academics would accept that there are a number of cultural characteristics which, when displayed within a company, will enable it to learn more effectively.

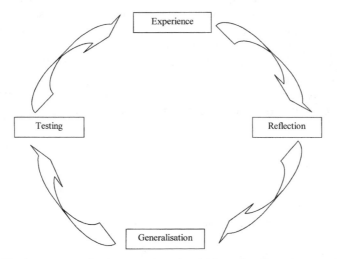

Figure 5.7 The learning cycle. Source: *Experiential Learning: Experience as the Source of Learning and Development* by Kolb, D. ©. Reprinted by permission of Pearson Education, Inc., Upper Saddle River, NJ.

- *A learning approach* will ensure that various aspects of business policy and strategy formation are structured as a learning process, allowing continual development and revision of the business plan.
- *Participative policy making* allows all company members to contribute to the majority of business policy decisions.
- *New communication systems* can be used to facilitate company processes and allow all employees company-wide access to information for their, and the company's, development.
- *Formative accounting and control* are used to meet the information needs of internal clients and engender a climate of individual and group responsibility.
- *Internal exchange*, i.e. interdepartmental and business unit collaboration, generates an internal customer and supplier ethic that creates professional market dealings in spheres of negotiating and contracting the provision of products and services.
- *Rewarding* flexibility, and communicating that fact, promotes innovation and initiative; regular, collective reviews are needed.
- *Structures* vary from company to company and change over time. A small company growing to medium or large size will encounter many structural changes both internally and externally; a culture of flexibility will aid future changes that may be required.
- *Environmental scanners* are people assigned to monitor the external environment continually and provide the core information on which the company can assess potential opportunities and threats, and adjust the corporate strategy accordingly.
- *Intercompany learning* creates a learning ethos within the company, and this message of competitive advantage is carried by sales personnel and the like to suppliers, customers and competitors.
- *A learning climate* is necessary to encourage and facilitate individual learning and development. This creates a tolerance of innovation, experimentation and mistakes and creates a culture of lifelong learning and continuous company improvement.
- *Self-development opportunities* should be available to all members of staff, allowing the individual the opportunity for continuing professional development (CPD).

Company organisation

The organogram in Figure 5.8 illustrates the basic company organisation. The standard states that companies should provide a detailed version showing names, positions, etc. It does not need to be an integral part of the environmental plan, but it can be referred to and kept in a supporting environmental documentation folder (EDF).

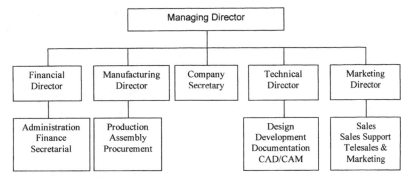

Figure 5.8 Organogram.

 👁 **Hint**

The standard does ask for names but it is accepted practice not to use employee names in your organisational chart. Any changes in personnel may well require frequent chart updates. Assessors will insist on inspecting this chart on a regular basis, so if you want to avoid unnecessary non-conformities use job titles, as opposed to names as chart changes will be less frequent.

EMS responsibilities

The EMS is managed by assigning roles and responsibilities to key personnel within the company. Examples of typical involved people and their responsibilities are as follows.

Managing director

The managing director has a number of different roles with environmental issues, including:

- overall responsibility for the effect that the company has on the environment;
- ensuring that the environmental management system is supported and maintained;
- ensuring that purchases made by the company are made with environmental policies in mind.

Environmental manager

Reporting to the managing director, the environmental manager has total authority for environmental management and has the full backing of management for actions that may be necessary in fulfilling this role.

The environmental manager would be responsible for:

- establishing and managing the environmental management system; this will be clearly documented and the documents will include details of the control processes within the company;
- ensuring that each department's responsibility for environmental issues is documented in an acceptable format and that adequate records are maintained for demonstration of conformance to these requirements;
- organising and scheduling internal audits of the environmental system to ensure continued adherence to documented requirements;
- identifying, recording and resolving, through designated channels, non-conformances within the environmental system;
- initiating appropriate action as needed and verifying that it has taken place;
- keeping abreast of relevant environmental legislative developments, issues and concerns;
- taking responsibility for communications to/from outside and inside sources of information;
- taking immediate action where necessary on receipt of communications requiring such action;
- day-to-day control of finances;
- preparing budgets, and short- and long-term planning;
- ensuring that funds are available to finance environmentally related projects;
- monitoring vendor performance in the supply of goods and services, in terms of environmental acceptability.

Production manager

The production manager would be responsible for:

- ensuring that plant and equipment are fully maintained and controlled under the preventive maintenance system;
- monitoring all maintenance schedules and emergency procedures with reference to environmental issues;
- controlling further activities following the development of a problem until any environmental deficiency has been corrected;
- taking whatever action is deemed appropriate in accordance with the EMS in emergency situations.

Procurement manager

The procurement manager would be responsible for:

- replying to any sales-related queries and ensuring that those concerning the environmental effect of the company's products are dealt with promptly;
- fuel oil delivery and aspects of waste disposal;
- ensuring that data sheets are obtained and are available for any products or substances, whether purchased or sold, that may have an environmental impact;
- submitting the environmental questionnaire to suppliers of products that may have an environmental impact, and for relaying responses to the environmental manager.

Environmental working group (EWG)

The EWG consists of the environmental manager and at least two co-opted members of staff. The EWG would be responsible for:

- co-ordinating environmental activities within the company;
- acting as a focal point for environmental matters;
- producing environmental progress newsletters for circulation within the company.

Employees

The employees would be responsible for:

- ensuring that operations are carried out in accordance with the procedures and work practices specified;
- reporting environmental procedure non-conformances to a responsible person specified in the EMS;
- suggesting EMS improvements to a responsible person.

Verification of resources

In a relatively small company, it is not possible to assign dedicated environmental posts. Consequently, the managing director should designate specific personnel who, in addition to their normal function within the company, have important roles within the EMS. These personnel, headed by a dedicated director or senior manager, form a small but highly motivated team that is responsible for overseeing the implementation and day-to-day management of the EMS. In addition, the team is responsible for carrying out the verification, auditing and monitoring of the EMS. The director or senior manager is responsible for ensuring that adequate personnel resources are available to provide full support for the requirements of the EMS and to carry out the necessary verification.

EMS communication and training

The environmental manager has the responsibility for communications, to and from external and internal sources and destinations, and the authority to take immediate action where necessary on the receipt of communications requiring such action.

Communications

The company should also establish procedures for receiving, documenting and responding to internal and external communications from interested parties regarding environmental effects and management. Depending on the communications received, decisions may be made regarding the following:

- Is an EMS non-conformance indicated?
  - If yes, what is the most appropriate action that should be taken?
- Is an environment management committee meeting required?

Details of environmental communications management should be kept separately in a designated environmental correspondence file.

Training

An induction and refresher training programme should be held to inform and educate personnel on the importance of complying with environmental policy, EMS objectives and preparedness for emergencies. Typical functions of the training programme include making personnel aware of:

- the environmental benefits of effective job performance;
- the potential consequences of departure from agreed operating procedures;

as well as:

- their role and responsibilities in compliance with EMS policies and procedures;
- their role and the correct actions to be taken in the event of an emergency situation.

Details of the application and management of environmental training are contained in the operational procedure.

Drills

Drills are carried out at random intervals to enable the response of the personnel involved to be assessed. Drills are the ideal means of testing, refining and improving emergency actions and procedures. To be effective, however, drills should be:

- planned carefully so as to provide the best simulation of a real event;
- monitored as they take place so that accurate performance records can be maintained;
- fully documented to permit later reference to events and actions;
- analysed when they are complete in order to identify possible improvements in both drill procedures and management of actual emergency situations.

Typical drill scenarios include:

- simulated spillage of a hazardous substance;
- fire;
- personnel contamination;
- release of toxic fumes;
- floods.

Drill monitoring

It is worth expanding on the importance of monitoring drill activities. The employee undertaking the drill will need to satisfy the drill criteria specified by the ISO 14001 standard. The close monitoring of the drill will also satisfy the environmental manager that the employee has achieved a particular standard and that further environmental training is not a requirement. Conversely, a less than satisfactory drill performance will signal the requirement for additional training.

The logging of the drill activities will provide evidence to the assessor that a drill has been undertaken to the required standard and that environmental training is ongoing. Often, drills will be conducted as part of an audit, but random, unexpected drills should be undertaken in order to simulate real-life situations.

Presenting the plan

The plan outlined above needs to be converted into a document at some point in order to be presented to senior management for approval, or possibly even to shareholders for acceptance. Whoever it is that needs to be convinced of the importance of implementing an environmental management system in the company will need some kind of document showing that it is a sound idea and will benefit the company. To that end, some guidance on the presentation of the document is provided.

When an environmental management plan is employed, a logical framework to develop and implement an environmental management system is used. This framework, which is outlined in this chapter, forms the basis for the remaining chapters in this text. The main subject areas are shown in Figure 5.1. Please note that this is not the only way to order the environmental management planning process; it is one logical and practical framework. As implementer, you should be aware that the content and prime focus of the environmental management plan will vary depending on what the company hopes to achieve, although the components and basic structure will remain constant. At least some of the following subheadings and information should be included before the body of the plan itself. A brief explanation is given below to justify their inclusion, but ultimately it will be you who will decide what subheadings are relevant and whether they are to be included in the plan.

Executive summary

The executive summary is usually the first section to be read and sometimes the only one accorded that favour. Take the opportunity here to summarise the environmental management plan so that the reader can see and understand the key points at a glance.

It is important to try to use the executive summary to gain the reader's interest. The text should be short, clear and concise, and should be used to, if not to sell the system, tempt the reader to find out more. It should, ideally, be no more than one page, and definitely no more than two pages, long. It is worth noting that if the first paragraph fails to grasp the attention of the reader then the rest of the text, even in the executive summary, has only a moderate to low chance of being read.

Title page

The title page of the environmental management plan should be clear and precise, and should have impact. You may well have produced many reports before, but remember that if this is a new concept to the company then you might want to give the whole document a second thought, beginning with the title page. If you are the extrovert type and are considering illustrating binders, folders and title pages with wild mushrooms, streams, trees and other environmental icons, that is fine as long as the message gets across. In the end, the EMS document must fit the style of the company and must not lose impact because of inappropriate presentation.

The title page should give a good impression because it is the first page that will be seen. It must contain the following information:

- the name and address of the business (and the registration number if appropriate);
- business telephone and fax numbers;
- the name and address of the managing director or proprietor;
- the names of the author and the manager responsible for environmental issues;
- the date (month and year) of writing, and the version number.

It is important that the title page is well presented. For example, you may decide to use your business or brand logo.

Contents page

An environmental management plan should be well presented and easy to read. Therefore, a list of contents and page references is necessary. Readers should be able to navigate the plan quickly and easily.

Environmental definitions

Try to keep environmental jargon to a minimum. Where it is unavoidable, try to ensure clear understanding by providing definitions of the terms used. If you are implementing ISO 14001 and are struggling for definitions, the ISO 14004 general guidelines document provides many useful ones, as does Appendix 1 of this book.

Company description

Here, a clear description of the company's business products, services and activities should be provided together with a history of the company and its overall corporate objectives. If there is already a business or marketing plan, a quick 'cut-and-paste' job should see this task completed swiftly. Key personnel who will be involved in the environmental management implementation process, i.e. the environmental working group, should be identified and their duties and responsibilities detailed. If it would be beneficial, include some detail of each person's skill set and how this will aid the implementation process.

6 Monitoring and measuring

An essential part of an effective environmental management plan is the assessment of how well, or how badly, the plan is working. This chapter provides an outline of the control and evaluation mechanisms that will help in the assessment or appraisal of current and previous environmental management plans.

Presenting an annual environmental management report provides an opportunity to review how well a company is achieving its targets and objectives, and assesses the way in which the environmental management planning function is intended to operate. The report functions as a monitoring and measuring device by providing a point of reference, or benchmark, against which to measure the progress that the company has made since the targets and objectives were set.

This chapter covers four main areas and is an attempt to offer clear explanation as to their use and operation.

- environmental plan review and control aims;
- control mechanisms;
- the evaluation process;
- the review process.

Environmental plan review and control aims

This section outlines the aims of the plan review and control processes. Four of the most common aims are:

- to allow problems or developments that do not match planned or budgeted schedule to be identified early and addressed if required;
- to identify the causes of problems or developments that do not match planned or budgeted schedule and act to nullify their effects;
- to provide input into the ongoing environmental management function of identifying environmental management aspects and impacts;
- to act as a performance indicator and stimulus for environmental management personnel.

Control mechanisms

In this section, describe the control mechanisms that the company has implemented in an attempt to satisfy the plan review and control aims. These mechanisms may include or involve the development of:

- performance criteria and standards;
- acceptable ranges within which these criteria and standards will be deemed to have been satisfied;
- procedures which will provide suitable and reliable measures of results;
- the means to compare the results achieved with the standards and criteria set;
- systems which enable effective corrective measures to be taken;
- a reliable means of forecasting outcomes.

The degree to which budgeted targets for each product within the corporate portfolio have been realised, measured by value and volume, should usually be checked month by month. A system – an existing budgetary system would suffice – must be established to enable the production of reliable and useful information as a matter of routine, as and when it is needed.

Data must be available at the appropriate level – aggregated figures showing the savings derived through implementation of the EMS, for example, detailed by department if required. Budget responsibilities need to be delineated clearly, e.g. a clear budget responsibility should be specified for the environmental management function and will encourage accountability.

Plan review meetings can also function as occasions on which certain kinds of information can be disseminated. For example, environmental management research data which may have been commissioned by one department may not easily 'trickle across' in less formal situations.

- Controls will also operate over new product development, with specific reviews of the development of such products or services.
- Continued shortfalls against projected performance should trigger revision or remedial activity.

You should now be in a position to evaluate the previous environmental management plans.

The evaluation process

Evaluation of previous environmental management plans

To regulate and control the environmental management plan effectively, it is necessary to use information to locate the causes of underperformance precisely, and to spot those things which are working and which the environmental management plan is therefore getting right.

Two main types of measure can be used: (1) environmental management costs to sales ratios and (2) customer tracking.

Environmental management costs to sales ratios

This ratio relates the amount spent on environmental management activities to the sales that have been achieved by the company. These are important, but broad, measures that must be interpreted carefully, and they provide a detailed check on environmental management expenditure. For example, acceptable annual environmental management costs could equate to 2 per cent of sales revenues.

Customer tracking

As with marketing and advertising initiatives, it may be useful to monitor how customers feel about the company's products and its environmental management activities. Such tracking includes a wide range of qualitative and quantitative measures of customer reactions taken from panel data, internal records of customer complaints, sales force reports, focus group interviews and surveys.

Because each business is different, it is difficult to outline a standard system for evaluating performance. However, the following need to be described:

- What has happened since the previous environmental management plan.
- How this accords with the time-scale or programme indicated in the previous environmental management plan.
- How it measures up to intended progress.
- Why this has happened.
- What extra costs (if any) have been incurred.
- How it fits into the budgeted figures.
- What actions are required as a result.

Of course, this may be the company's first environmental management plan and there may, therefore, be little or nothing to compare it with. If the company does have a previous environmental plan, a simple review would assess whether previously specified targets and objectives had been achieved. If they have, set some new ones; if they have not, find out why not. The ethos of continuous improvement on which the ISO 14001 standard is based must be evident within environmental plans. That said, do not make them a huge burden for the company to achieve.

Environmental management procedures (EMPs)

All of the procedures in the environmental management plan are written to safeguard the effective operation of the management system. Of the two sets of procedures, the operational procedures put into action processes that will eventually allow the realisation of the objectives and targets. The management procedures, those explained in this chapter, provide the controls. They ensure that the environmental management system is monitored continually and, where necessary, corrected and improved over time.

The following environmental management procedures conform to ISO 14001 requirements and offer possible bases for procedure formation. As always, they can be used as they are presented here or modified to suit the company's requirements.

The five major environmental management procedures are as follows:

- control of non-conformities;
- management review;
- corrective action;
- document and data control;
- internal audit.

As specified in other procedural guidelines, the start of any procedural prescription should specify the purpose of the procedure, its scope, the person responsible for its execution and its administrative process.

Management procedure – control of non-conformities

Purpose

This procedure describes how non-conformities are controlled and reported (see Figure 6.1).

Scope

This procedure establishes the way in which non-conformities are classified, recorded and evaluated.

Responsibility

The environmental manager is responsible for documenting non-conformities, presenting them at management review and ensuring that follow-up actions are undertaken.

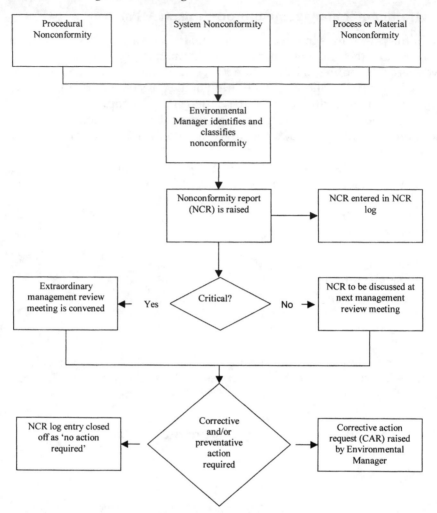

Figure 6.1 Control of non-conformities.

Procedure

There are three types of non-conformity:

1   Procedural non-conformity – when personnel have failed to follow environmental operating procedures (EOPs) or other forms of instruction contained within the EMS (typically identified during internal audit or periodic review by the certifying agency), or when the non-conformity is an accident or a hazardous occurrence. The most obvious example is where someone has failed to complete a new product introduction (NPI) form when introducing a new product into the company.

2   System non-conformity – when a fault or inadequacy is identified in the EMS and change is required to assure the safety and protection of the internal and external environment. Using the NPI example again, the procedure for introducing new products into the company may not have been created.

3   Process or material non-conformity – an accident or occurrence not related to either of the above types of non-conformity (typically involving an unforeseen breakdown or failure of equipment, process or facilities). An example of this type of non-conformity may occur when a piece of machinery fails or an oil or diesel spillage creates a potentially hazardous situation from the threat of fire or of water and / or land contamination.

Non-conformity identification

Non-conformities may be identified by any company employee, or by anyone, such as a subcontractor, who may be working for the company.

Non-conformity reporting

The non-conformity reporting chain is achieved through the use of existing management structures and the established interface.

It may not be feasible or practicable for initial reporting of a non-conformity to be carried out in writing. In many cases, a verbal report will be made to the environmental manager, who will carry out an initial investigation and then produce a written report. Unless it is impossible, all reports should be put in writing eventually, regardless of the initial method of transmission.

All final non-conformity reports submitted to the environmental manager should conform to the reporting form layout distributed by the environmental manager and should include details of immediate actions taken, or to be taken.

A non-conformity can be either critical or non-critical:

- A critical non-conformity has a direct and immediate effect on safety or protection of the environment. For example, a faulty piece of equipment might be identified or an environmental operating procedure (EOP) or other form of instruction might be found to contain an error that could affect the environment.
- Any other type of non-conformity is, by definition, non-critical.

The environmental manager has the task of deciding whether a non-conformity is critical or non-critical.

Environmental manager actions

When details of a non-conformity are received, the environmental manager should produce a non-conformity report (NCR) in which:

- the type of non-conformity is identified;
- the non-conformity is determined as critical or non-critical;
- the non-conformity is described in sufficient detail to allow management to identify the appropriate corrective action.

Every NCR should have an issue date and be identified, for instance, by a four-digit code, e.g. 0001. If the company has a preferred numbering system, it should be used or a new one created.

The environmental manager should retain copies of all NCRs and maintain a master log of them. The log identifies NCRs by issue number and date, and shows the date of the management review meeting at which the close-out actions and responsibilities were discussed. Each NCR is closed out in the log by the allocation of a corrective action request (CAR) number or numbers or when management review decides that no action is to be taken.

If the non-conformity is designated as critical, the environmental manager will call an extraordinary management review meeting as soon as possible. Temporary remedial action may also be taken immediately and detailed on the NCR. If the non-conformity is non-critical, the NCR will be presented at the next scheduled management review meeting.

Management procedure – corrective action

Purpose

This procedure describes how corrective and preventive action is initiated and maintained within the EMS (see Figure 6.2).

Scope

This procedure establishes the way in which corrective and preventive action is documented and reviewed.

Responsibility

The environmental manager is responsible for producing a corrective action request (CAR) for any corrective action decided at a management review.

Procedure

The CAR will provide brief details of the non-conformity to which the

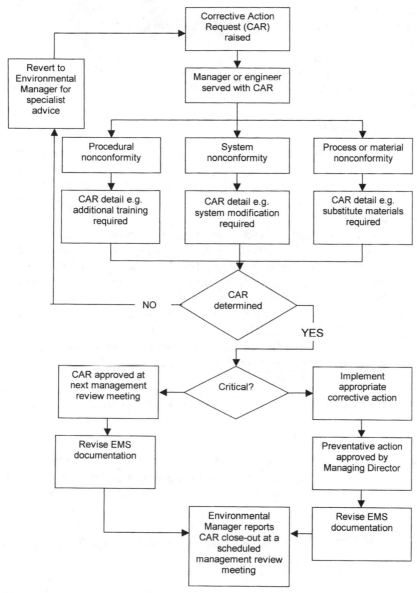

Figure 6.2 Corrective action process.

corrective action relates and will provide information on what needs to be done. It identifies the addressee – the person required to complete the corrective action – and includes a time-scale for completion of the corrective action.

The environmental manager retains copies of all CARs issued and maintains a log of them which shows:

- CAR number;
- non-conformity report (NCR) relating to the CAR;
- addressee;
- date CAR was raised;
- CAR status (actioned or to be actioned);
- CAR close-out date.

Close-out of corrective actions

CARs relating to a procedural non-conformity are closed out when the auditor who raised the NCR has reviewed the corrective action and accepted that it addresses the requirements of the NCR. If the original auditor is not available, the environmental manager may close out a procedural non-conformity.

CARs relating to cases of system or material/process non-conformity are closed out when the environmental manager has reviewed the corrective action and accepted that it satisfies the requirements of the NCR.

The environmental manager will produce a CAR close-out report on all closed-out CARs. This report will briefly summarise the NCR, the corrective action required and the actual action taken. The environmental manager will retain copies of all CAR close-out reports.

Copies of CAR close-out reports will be distributed to management at the next scheduled management review meeting.

Management procedure – document and data control

Purpose

This procedure describes the control of documents and means of electronic data control (including indexing, data storage and protection, and archiving) that are essential to the effective operation of the EMS (see Figure 6.3).

Scope

This procedure should apply to the following:

- environmental management procedures (EMPs);
- environmental operating procedures (EOPs);
- forms and reports;
- work or other instructions;
- any other environmental documentation.

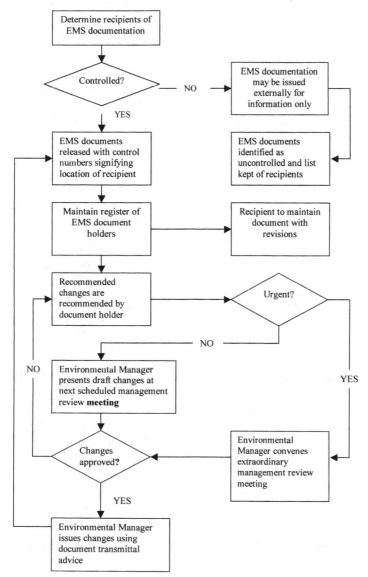

Figure 6.3 Document and data control.

Responsibility

The issue, amendment, approval and distribution of all EMS document-ation should be carried out under the direction and authority of the environmental manager. Document holders are responsible for the maintenance of controlled documentation and for ensuring that all obsolete documentation is returned as instructed.

Procedure

All documents relevant to the EMS should be reviewed, maintained and controlled to ensure that:

- documents and their subsequent revisions are reviewed and approved by the environmental manager before use;
- the issue and revision status of the documented EMS is known to all users through the use of controlled documents;
- persons issued with controlled documents are aware of their responsibility for maintaining the revision records;
- up-to-date issues of appropriate documents are provided at any locations where critical operations are performed;
- all copies of obsolete and superseded documents are withdrawn from use to be destroyed or archived;
- any obsolete documents retained as records are identified suitably and that a master list is maintained of all documents and data under control.

Control of company-originated EMS documentation

All such documents should bear a control header design (as used in this procedure) that includes (in designated form and location):

- a corporate identifier (as appropriate), document title, chapter number and title;
- issue status such as draft or the date (expressed as month and year);
- revision number, starting at 00 and increasing sequentially in units;
- page number, or chapter number and page number (expressed as x or x-xx) or page number and total number of pages per chapter (expressed as x/xx); and
- forms that have a unique numbering sequence which indicates their function.

Distribution matrix

The environmental manager should maintain a distribution matrix that records the following information about all documentation relevant to the EMS:

- document title;
- controlled copy number;
- location/holder;
- revision status;
- issue date.

Any changes affecting the location and/or status of individual publications or copies should be recorded on the distribution matrix.

Document transmittal advice

When a publication or document is issued, or if the environmental manager issues a revision, it should be accompanied by a document transmittal advice form. This form is addressed to the person who will be receiving the new publication or who is responsible for revising an existing publication.

If the form accompanies a new publication, the recipient is required to signify receipt and respond in accordance with the instruction given on the form. The revision record should be completed and signed in accordance with the document transmittal advice form instructions, which should also detail the response required – see Chapter 9 for a sample document.

Revision record

Other than forms and minor documents, a revision record prefaces company-originated EMS publications. This is a record of revisions to the specific publication and, therefore, provides essential evidence that a publication is current. It is the responsibility of the person having charge of the publication to ensure that the record is maintained correctly and accurately.

Control of electronic data

All electronic data that originate from within the company must be identified, indexed and stored in accordance with current company instructions. Access to company computer systems should be granted only to those persons authorised. Access to computer files should be similarly controlled. Such precautions can minimise the possibility of accidental or deliberate corruption of essential information, or electronic infestation that might corrupt or destroy data.

Control of documents from external sources

The environmental manager should be required to maintain an up-to-date list of all externally supplied publications and should also be responsible for the purchase and supply of such publications. Each publication should be given a unique number or code upon receipt. The date of issue should be clearly stated. Obsolete copies should be withdrawn from service and 'SUPERSEDED' marked prominently on the front cover.

Management procedure – internal audit

Purpose

This procedure describes how internal audits of the environmental management system are scheduled, planned, carried out and recorded.

Scope

This procedure establishes the conduct by which internal audits are undertaken.

Responsibility

The EMS is subject to regular audit to ensure that it continues to meet its objectives and that personnel are complying with procedures and work instructions.

The environmental manager is responsible for scheduling and documenting audits. All company personnel are responsible for assisting the audit process by participating and assigning resources as required.

Procedure

The environmental manager should draw up an audit schedule that defines the number, frequency and subjects of audits of the EMS. This schedule will take account of work scheduling and any other critical activities, and will include, as a minimum, at least one audit of the EMS every year. The environmental manager should nominate a suitably qualified person to act as auditor and should make known to all employees the name and location of the auditor on company noticeboards. The completed audit schedule will need to be approved at a management review before it can be put into action.

Before the audit, the auditor will decide what particular aspects of the EMS are to be audited and what objective evidence will be sought. All persons and/or departments to be audited will be notified at least 1 week in advance of the planned date of the audit. The subject of the audit – i.e. the particular procedure(s) or work instruction(s) to be covered – and any personnel with whom the auditor wishes to talk will be identified clearly.

During the audit, the auditor will examine objective evidence to confirm that operations are being carried out as required by procedures and work instructions. The auditor will also discuss procedures and work instructions with personnel and evaluate any corrective action taken in relation to CARs with a view to authorising close-out. Following the audit, the auditor will produce an audit report, which will include:

- date, time and location of the audit;
- objective of the audit;
- basis of the audit (those procedures and work instructions that were verified);
- details of personnel audited;
- details (including NCR numbers) of any non-conformity noted;
- details (including CAR numbers) of any corrective action evaluated.

If non-conformities are found, a follow-up audit may be conducted to ensure that corrective action has been successful. Alternatively, the corrective action may be reviewed at the next scheduled audit. The environmental manager is responsible for deciding whether a follow-up audit is required, but one will be required and conducted for all critical non-conformities.

Copies of the audit report will be provided to the auditee(s) and to all members of senior management involved in management review. The environmental manager will retain copies of reports, and completed audits will be discussed at the next scheduled management review meeting.

Management procedure – management review

Purpose

This procedure describes how management should review the EMS at regular intervals to ensure that it continues to satisfy the environmental policy.

Scope

This procedure establishes the conduct by which a management review is undertaken.

Responsibility

The managing director is responsible for ensuring that the EMS is reviewed according to this procedure.

Procedure

The review should be conducted by means of a preplanned meeting chaired by the managing director and attended by the environmental manager and one (more if required) member of the environmental working group.

Management review meetings should take place at least once in each 6-month period. These meetings can be held annually depending on the company's circumstances, but every 6 months is recommended. The environmental manager should advise the managing director when to announce the precise date of the meeting.

Unscheduled, extraordinary review meetings may also be held should circumstances dictate, particularly in the event of fire and/or floods etc.

The environmental manager is responsible for:

- convening a management review meeting and drafting an agenda for approval by the managing director;
- maintaining the environmental records that provide input to the management review;
- ensuring that minutes of the review and all agreed actions are recorded for subsequent approval by the managing director;
- preparing and updating an EMS audit schedule for approval by the managing director.

The environmental manager should maintain an EMS audit schedule showing:

- the frequency and timing of audits;
- specific areas and activities to be audited;
- identity and qualifications of auditor(s);
- auditing and reporting criteria.

The EMS audit schedule should be regularly drafted and/or updated by the environmental manager for approval by the managing director. The redrafting should be carried out at least once in any 12-month period. Every company employee is responsible for compliance with the EMS audit schedule by participation, assigning resources, etc., as required.

The management review agenda typically includes:

- review of the minutes of the previous meeting and actions carried forward requiring close-out;
- results of EMS audit activity;
- review of the environmental manager's summary report of non-conformities since the previous meeting;
- review of organisational management procedures;
- review of administrative procedures;
- review of environmental operating procedures;
- review of personnel responsibilities and authority;
- review of documentation and record keeping;
- review of and adherence to EMS policies, procedures and instructions;
- need for additional familiarisation or on-the-job training;
- results arising from analysis of any critical non-conformity such as personal injury, equipment damage or pollution incident;

- corrective action request close-out reports produced on non-critical and critical non-conformities since the previous meeting;
- corrective action taken on operational defects or procedural amendment in the EMS and further measures to improve its effectiveness;
- recommendations from employees and EWG meetings for measures to improve EMS effectiveness;
- the degree to which the environmental policy continues to meet the company's current objectives and statutory compliance requirements.

The environmental manager should revise the EMS audit schedule according to the actions arising from the management review and present it to the managing director for approval. The results of the review should be brought to the attention of those persons responsible for implementing the changes proposed. Ensure that all actions arising from a management review are closed out and signed off by the managing director.

Continuous improvement

Monitoring and measuring the existing environmental management plan is a necessary requirement for ISO 14001. A similar requirement for the standard is a programme or system that facilitates continual environmental improvement. It is too simplistic to set a number of targets and objectives, achieve them, and sit back and smoke a cigar. Once existing targets and objectives have been achieved, more need to be set. These future targets and objectives may take the form of new targets based on existing objectives, for example a further 5 per cent saving on energy usage, or completely new objectives and targets.

As mentioned at the outset, do not be too adventurous with energy savings or waste minimisation levels each year; set achievable annual goals. It is better to have achieved a target of a 5 per cent saving each year for 5 years than a 25 per cent 1-year target. The latter may well be difficult to achieve in 1 year, and yet failure to do so may have a negative effect on morale and the programme of continuous improvement.

7 EMS audit and management review

Audits – general

When all of the management and operational procedures have been implemented, it is time to test them to see whether they work and whether they are being used correctly. This chapter is designed to set out the audit procedure to test the environmental management system and to establish a management review format to determine further development or changes to the system. The environmental management system (EMS) audit will help to determine whether:

- the business activities are conforming to the requirements of the EMS;
- there is employee awareness, together with procedural familiarity and compliance;
- there is operational relevance, accuracy and effectiveness of the environmental operating procedures (EOPs) and environmental management procedures (EMPs);
- there is proper determination of the adequacy of the EMS by senior management.

Audit plan and procedure

If a quality auditing plan and procedure is currently in use, adapt it. It will not take a lot of adapting but it will save a lot of time. The way in which the EMS audit will be carried out will have already been specified in one of the environmental management procedures; if you are also an auditor, you will know that this procedure itself will also be audited, so make sure you follow it.

The ways in which internal audits are scheduled, planned, carried out and recorded are detailed below. It is the environmental manager's responsibility, together with that of the working group, to develop an audit plan and schedule. A suitably qualified auditor should undertake the audit. If one of the working group members has auditing experience, take advantage of this; if not, someone who has the relevant skills should be co-opted.

Audits, of whatever kind or application, have received some bad press over the years. It may be the inspection, checking or investigation of the activities of an individual, department or company by complete strangers that brings the word 'caution' to mind. If it helps, add a cautionary line to the audit notification to employees that emphasises that it is the process that is being tested and not the individual.

By the time the system is audited, 3 months will have passed since the company started using the EMS procedures. During that time, the environmental management system will have been fully operational and will have been used by every member of the company according to the procedures. The time set has not been plucked from the air, it is a minimum requirement and it is that specified by the ISO 14000 standard.

Audit plan

This final system audit requires that an audit of the complete system is undertaken that covers all operations of the company. It is best to start by devising a schedule of all the audits that have to be undertaken. Your audit schedule should show the audit functions to be undertaken. It is beneficial to spread them throughout the year at a frequency of either one or two per month. Some audits are more time-consuming than others, so the schedule should be planned carefully. If it is necessary to have two audits in one month, try to ensure that they are within the same department or are relatively small or similar in nature.

Remember, this is your plan; plan it as you wish, colour-code the audits if this helps the planning process. Schedule them as you will, as long as they are all completed within 1 year. By far the most time-consuming of all of the audits is the system audit. The system audit takes place annually, and it may be useful to identify this particular audit with a different colour; this will highlight its significance to all those involved, including the assessor (for further details, see Chapter 9).

During the implementation phase of ISO 14001, there will be many temptations to do or not to do something. Most of them will be of an ethical nature. Some will be considered small and some will be considered huge. Most of them may have some degree of justification, depending on your level of fatigue or frustration. Remember that there are trade-offs in business, so be flexible; it may be necessary to lose the odd battle to win the war.

One example of a trade-off may be the acceptance of low prioritisation, or late response, by some departments when actioning environmental procedures because of operational requirements. It may be that a late response is preferable to no response.

One temptation that should be avoided, however, is to omit auditing some elements of the environmental management system. During progress through the ISO 14001 process, it will very quickly become clear that the assessment team could not hope to assess the whole system during the 2 days (depending on company size) that will be allocated for the final accreditation. They cannot do a full audit on the whole system in that time. They know it. You will know it. And they will know you know it. So they sample audit. This is a sensible option. The assessors save time, the company saves money. The lead assessor will select certain areas of the company to assess and the assessment team will be allocated its duties within these areas.

This warning is not a betrayal of any confidences. When the company receives any assessment, interim or final, the assessor will specify that the assessment has been based on a sample of the system. If the company passes the assessment, the assessor will also add comment such that it has passed the assessment based on those parts of the system assessed. The assessor will mention that non-conformities may still exist, but in those areas that were not assessed, and that these may be found on the next visit.

To audit your EMS selectively would be something akin to sitting O levels at school and only swotting the questions that you thought were most likely to come up in the examination. This strategy is risky and unprofessional, and can lead to a considerable amount of wasted time and money – and a failing of the examination. In short, the whole system should be audited. In fact, it is far simpler to audit the whole system. If you feel that 3 months is insufficient time to test the system fully, extend the time. If you are selective, you may forget what has been audited and what has not and, if you miss something, be assured that the assessor will not.

Audit notification

Under the requirements of the ISO 14001 standard, it is necessary to carry out regular audits of the environmental management system to ensure that it continues to meet its objectives and that personnel are complying with management and operational procedures. Before any audit is undertaken, each person to be audited must be notified a minimum of 1 week in advance of the audit date. It may also be useful to post a list of all employees to be audited, including times and dates, on the company noticeboards. The notification of audits may seem to be a minor piece of auditing etiquette, but beware: the assessor will be looking for proof of audit notifications, and a non-conformity will result if they are not found.

At this stage of the implementation process, you may find that you briefly enter the twilight zone of environmental management. The dissemination of audit notifications strangely coincides with the mass application for holidays, dental appointments and sabbaticals for missionary work in deepest Africa (Figure 7.1).

The audit notification requirements can be very simple in format but must provide advance notification of the following:

Figure 7.1 Audit escapee I presume?

- who will be audited;
- when and where the audit will take place;
- details of the topics to be covered by the audit.

In addition to the above requirements, a brief note as to how each person is to answer the audit questions and where the answers can be located will also prove to be time well spent. It is always helpful to spell out the audit requirements clearly and in full. Audits can at times be viewed with suspicion and trepidation, so time taken to reassure will aid the process and improve results.

 **Hint**

Set a completion date for all persons to be audited. This will ensure that all audits are completed quickly. Audit times may not suit everyone and some will change. Having a final completion date will help to ensure that any time changes remain within the allotted time period.

Audit questions

An example of some audit questions that might be asked is given in Figure 7.2. The presentation is simple, with space available for answers. Prior notification of any audit allows everyone being audited the opportunity to dash to the nearest environmental manual and familiarise themselves with the relevant procedures.

Question	Answer
1 Who is responsible for the identification, handling and storage of all waste?	
2 What procedure defines the methods that shall be used for the handling, storage and disposal of liquid and solid waste produced in the office, workshop and stores areas of the company?	
3 Who is responsible for documenting and managing the environmental management system?	
4 Where would you find information regarding the way in which spillages of hazardous materials should be dealt with?	
5 Who is responsible for the storage of all materials?	
6 What is the name of the plan that is designed to be a dynamic and open-ended list of objectives and goals for the company to achieve in accordance with ISO 14001?	
7 What is the correct method for the disposal of waste categorized as rubbish?	
8 Who is responsible for the storage and maintenance of all records and completed forms that are part of the EMS?	
9 Who is responsible for ensuring that personnel are adequately trained and are competent before carrying out any process or operation?	
10 What environmental documentation is required following the introduction of a new process, product or substance within the company?	

Figure 7.2 Typical audit questions.

When setting the audit questions, choose whether to make them difficult or simple. The benefit of the simple option is that people will feel less threatened by the task and more willing to be audited, and will be more inclined to take time to read the relevant procedures. The audit can be simplified by prewarning the individual of the systems and procedures that you intend to audit. Because of operational pressures, most employees will appreciate more direction and less reading of procedures. This does not diminish the exercise. Do not forget that this is likely to be a new system and a new learning experience for everyone. It is best not to push too hard at this stage, with time and use everyone will come to understand what is required of them and how the new system works.

Constructing the audit report

Having completed the audit, as with everything else, evidence is required. The construction of an audit report is the next task. This will provide you and the assessor with two pieces of evidence of a successfully working procedure and system. The first is the proof that the audit has been carried out in accordance with the auditing procedure. The second will be the proof that those employees audited have read the procedures and are at least familiar with the general requirements of their role in the environmental management system.

The audit report should consist of three elements:

- a cover sheet;
- the response to the audit questions;
- the conclusions of the auditor.

An example of an audit report can be seen in Chapter 9.

The audit is, in effect, the company's annual medical check-up. During the audit, the auditor will examine:

- the organisational structure;
- management and operational procedures;
- the workplace, including layout and operations;
- how well the EMS meets the requirements of the company's environmental objectives.

You may find that it is hard to fault what you have done, particularly if, as part of the environmental working group, you have been both implementing and auditing the procedures. When auditing the system, it is a great temptation to correct an error that is found but not record it in the audit. You will find sometimes that the last thing you want to do is to raise a non-conformity, and all the associated paperwork, over something trivial.

Someone once said, 'you can't make a good omelette without breaking some eggs'. Similarly, it is not possible to audit an environmental management system successfully without incurring some non-conformity. The more non-conformities that are found the better the system will be, and the more satisfied the assessor. First, you will have demonstrated that the system has been audited effectively; second, and more importantly, you will have demonstrated the working of the management procedure by identifying, raising and actioning non-conformities.

The report

The audit report (Figure 7.3) contains details of individual environmental audits that have been conducted in accordance with the requirements of the EMS auditing procedures. As mentioned earlier, before final assessment for ISO 14001 it is necessary to operate the EMS for 3 months. At the end of the 3 months, the audit process detailed above goes into action. The final piece in the ISO 14001 jigsaw is to review the findings of the audit. The audit may have raised a number of non-conformities about certain procedures, or indeed the operation of the system itself. A management review is a system requirement and another procedure that will be assessed.

Review by management

According to management review procedures, the management review will be constructed and organised by you – the environmental manager – or by members of the working group, but only your senior management team can carry out the review itself. The format of the management review will be similar to:

- notification of management review;
- management review agenda;
- management review;
- management review minutes.

The management review procedure will require the EMS and the audit results of the system to be reviewed at regular intervals to ensure that it continues to satisfy the requirements of achieving the objectives and targets

Auditee: Auditor:

Audit date: Audit location:

Audit objective: Audit report number:

	Yes	No	
NCR			NCR number:
CAR			CAR number:
Follow-up action required? (see narrative)			Reference number:

Narrative

Signed (auditor):   Date:

Figure 7.3 Audit report.

of the company and the procedural requirements for the ISO 14001 standard.

Management review agenda

At the management review, examine the minutes of the previous meeting and identify actions carried forward that need to be closed out (Figure 7.4).The agenda detailed below can be regarded as typical for a routine management review meeting.

- Results of EMS audit activity.
- Review of the environmental manager's summary report of any non-conformity since the previous meeting.
- Review of environmental management procedures.
- Review of environmental operating procedures.
- Review of personnel responsibilities and authority.
- Review of documentation and record keeping.
- Need for additional familiarisation or training.
- Review of training syllabuses.
- Analysis of results of critical non-conformities such as personal injury, equipment damage or pollution incident.
- Corrective action request (CAR) produced on non-critical and critical non-conformities since previous meeting.
- Corrective action taken on defects or procedural amendment in the EMS and further measures to improve its effectiveness.
- Recommendations from employees and environmental working group meetings for measures to improve EMS effectiveness.
- The degree to which the corporate environmental policy continues to meet the company's current objectives and statutory compliance requirements.

Agenda approved: Date:

Extraordinary management review meeting

If a critical non-conformity has been identified, i.e. a non-conformity that has an immediate and direct effect on safety or protection of the environment, a responsible person should take immediate preventive action. Following consultation with the senior manager or director, an extraordinary management review meeting may be convened in order to approve further corrective action.

Each non-conformity is evaluated at a management review meeting, where it is decided whether corrective action is required or any prior action taken has achieved this aim. The management review process also examines the EMS as a whole to ensure that it continues to meet the objectives of the corporate environmental policy.

Date/time:

Location:

Present:

Item	Topic	Action
01	Initial meeting. Consequently, there are no previous minutes to review and no previous actions have been carried forward.	
02	Reported on personnel audit progress and discussed the proposed audit schedule.	
03	The two NCRs that had been submitted were discussed. CAR action had been initiated.	
04	The use and effectiveness of EMPs was discussed. No problems or difficulties arose.	
05	The use and effectiveness of EOPs was discussed. No problems or difficulties arose.	
06	Personnel responsibilities were discussed and considered adequate. No points arose.	
07	Documentation and record-keeping were discussed. Current standards were considered adequate, and no points arose.	
08	The need for additional familiarisation and training was discussed. Current levels were considered adequate. No points arose.	
09	There have been no critical non-conformities raised to date. No points arose.	
10	The two current CARs were discussed. Recommended actions will be initiated by the environmental manager.	
11	No employee or working group recommendations or suggestions have been received.	
12	The functionality of the environmental plan was discussed and it was agreed that no changes were required at this stage.	

Figure 7.4 Environmental management review minutes.

Corrective action

Corrective action is the action taken by a responsible person under the direction of a senior manager to prevent the recurrence of a non-conformity or to improve the EMS.

Control of non-conformities

Any person within the company can, and should, report an actual or potential environmental non-conformity. All non-conformities should be reported directly to a senior manager, either in person or in writing. The senior manager will discuss the report with the originator and will assist in the accurate recording of the non-conformity.

Document and data control

Corrective action is formalised by:

- revision to the EMS documentation;
- issuing the changed documentation;
- initiating a follow-up audit to verify implementation and effectiveness of the corrective action;
- internal audit.

The cycle of activities is completed when the corrective actions are closed out under the internal audit procedure (Figure 7.5).

The role of the senior manager in the cycle of non-conformity reporting, the implementation of corrective action, follow-up audit and document control is essential to the effective operation of an environmental management system – see Chapter 6 for more details on corrective actions and control of non-conformities.

Management review

The senior management team should carry out an annual review of the EMS in order to ensure its continuing suitability and effectiveness. The reviews should establish the need, if any, to change policy, procedures, controls, objectives or other relevant matters, taking account of the audit results and changing circumstances and including legislation and the need for continual improvement.

Corporate environmental policy statement

The existing policy statement needs to be reviewed annually by the managing director, who should update the contents in order to stress the company's commitment towards continuous environmental improvement.

EMS Structure

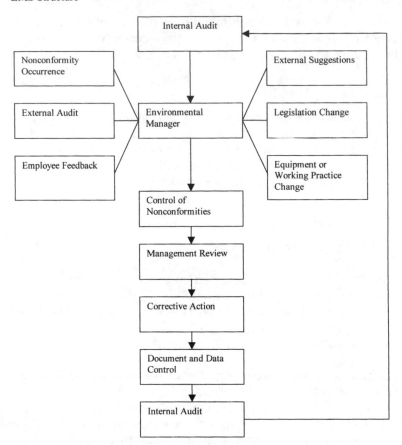

Figure 7.5 EMS structure.

Environmental working group

A major outcome of your initial environmental review should be the formation of an environmental working group (EWG), headed by a senior manager or director. The EWG should be tasked with meeting at least once a month in order to discuss any matters that have arisen and to progress any on-going environmental topics. Meetings can be arranged more frequently if the need arises, particularly if the company plans to introduce the EMS quickly.

Employee feedback

Employee involvement in environmental matters is fundamental to the adoption and maintenance of a successful EMS. A number of ways in which employee involvement in the EMS could be improved should be proposed, including a survey for suggestions as to how the company's operations could be improved in terms of environmental protection. In addition, all employees should be urged to highlight any process or operation that could have any adverse environmental impact.

Register of regulations

It is essential that a dedicated member of staff is fully aware of the regulations that apply to the company; consequently, a fundamental part of the planning and implementation process is the creation and on-going maintenance of a register of legislation (see Chapter 9).

Register of environmental aspects and impacts

A register of the environmental effects of normal and abnormal operating procedures, incidents, accidents and potential emergency situations should be maintained. This will assist in the monitoring of the company's effects on the environment and the development of any necessary remedial actions and procedures. The register of environmental aspects is explained further in Chapter 9.

The environmental management plan

An environmental management plan should be easy to read and avoid jargon wherever possible. Remember that most of the people reading the plan will not be environmental management experts. If you use complicated terminology, you should also include definitions and explanatory notes. In some cases, a glossary of terms is necessary and should be included as a separate section at the end of the environmental management plan (see Chapter 5).

Each section and page should be numbered so that the reader can easily navigate the plan. Visual aids such as tables, charts and diagrams are often the best way to convey complex information. Photographs of products, processes and business activities can be used to give life to your environmental management plan. Be candid and honest when conducting forecasts and writing the environmental management plan budget. Overoptimism can cause doubt in the reader's mind about the credibility and judgement of the author.

After the first draft of your environmental management plan has been completed, cross-check it for inconsistencies. For example, make sure that the environmental management aspects and impacts and environmental management procedures are in line with your company's objectives and targets. At this stage, you should also check that the content of the environmental management plan is relevant to the reader and discard any surplus material. It is also vital that the objectives are communicated and, wherever possible, agreed with those who are charged with the responsibility of achieving them.

Typically, companies have operational objectives affecting many areas of the company, any one of which can affect the selection of environmental management objectives and strategies. A company's business objectives are likely to encompass more than simply finance and involve those of both a qualitative and quantitative nature. Some of the more common objectives for many organisations include:

Qualitative	*Quantitative*
Market standing/reputation	Profitability through materials recycling
Innovation	Greater production efficiency
Management performance	Profitability through energy efficiency
Public responsibility	Identification of new products
Organisational development	Efficient operational processes

When framing the objectives, make it clear how they will be achieved. For example, an objective to increase market share with a low product price may be written as:

> Our objective is to decrease energy consumption in the coming year by 10 per cent by conducting an extensive energy usage analysis and supplier cost comparison.

You can set as many objectives as you like, but make it clear that they are relevant to the purpose of the plan and also that they are in line with the financial projections and all other aspects of the environmental plan.

The audit of environmental management objectives, policies and activities

From time to time, the company should conduct a complete review of environmental management objectives, policies and activities on a company-wide basis. Such a review is perhaps the most comprehensive approach to evaluating environmental management effectiveness. The review should aim to examine and evaluate the success or otherwise of the environmental management objectives and policies that have been guiding the company. This is a comprehensive review of both the activities of the company in relation to environmental management and also the environmental factors that are likely to bear upon the success, or failure, of the activities in achieving the objectives of the environmental management plan.

Environmental management plan budget

Budgets and resources are allocated in an attempt to meet business and environmental management objectives and to make the overall environmental management plan successful. Products performing particularly well or particularly poorly, from an environmental perspective, will obviously require quite different levels of activity, and, consequently, different allocation of budgetary resources.

Often, the costs involved in the environmental management plan are a very significant part of the overall operating costs of a company's operation. It is sensible to plan these costs systematically, to attempt to forecast their effect, and to keep as close control as possible over the way in which they are used.

Keep the budget section brief but comprehensive. The environmental management budget can be as comprehensive as the financial plan that you would write as part of a business plan. Many of the items that you may want to consider, however, are identical, and may be transferred directly from the business plan or profit-and-loss forecast.

The main tasks that you will have to conduct to develop an environmental management plan budget are:

- estimating total revenue and costs;
- allocating environmental management resources within the portfolio of products/services and between elements of the environmental management objectives.

Although there are a number of different ways in which the environmental management plan budget can be calculated, the environmental management plan follows a stepwise procedure in which each of these tasks is conducted in turn.

Estimating total revenue and costs

As the environmental manager, you will probably have a profit-and-loss forecast for your department or business, as will the company's bank manager or accountant. A profit-and-loss forecast is useful in showing what remains after all the money that enters a company has been spent. The profit-and-loss forecast has to be produced before the remainder of the environmental management plan can be prepared. Indeed, if there is not an acceptable net profit for any product at the end of this stage, then you may need to go back over the assumptions made elsewhere in the plan. In this way, the actual figures will determine how the business is run. The first stage of the environmental management plan budget involves using this information to calculate total revenue and costs as well as gross and net profit.

There are four variables that you need to know:

- revenue from customers (sales);
- costs of product or process (direct costs);
- all other non-environmental management overheads;
- environmental management costs.

Direct costs

Direct costs are expenses that are directly attributable to the procurement or manufacture of the company's products. They include:

- hire of factory plant and equipment;
- salaries of production-related people (factory workers, production manager, etc.);
- manufacturing utility costs (gas, electricity, water);
- distribution or transportation costs;
- wages costs (if you are selling people's time).

All other non-environmental management overheads

Any expenses not directly attributable to the sale of an individual product are called overheads. For example, a telephone and fax machine will be used in your department or business and the company may employ a receptionist. Although your department or business may not function properly without any of these things, all of them are classified as overheads because they are not directly needed to manufacture the products sold by the company. Some costs are extremely hard to define and may need to be split between direct costs and overhead classifications. Other examples of environmental management overhead include:

- management salaries;
- office rent and rates;
- insurance;
- leasing and rental of office equipment.

When calculating costs, consider simultaneously the decisions that you make with regard to the level of company commitment, as this will directly influence the total environmental management budget.

Allocation of the environmental management budget

The environmental management budget has to be allocated to the various departments or elements of the environmental management plan. New products, or products that the company is attempting to reposition environmentally, may require a higher environmental management budget commitment than others. Estimates of additional revenue for new environmentally friendly products or by-products should help to make the allocation decisions a little easier.

The costs of distribution and production will have been accounted for under the estimates of total costs; however, budgets for promotion, environmental management research and new product development will still need to be calculated. When these budgets are set, the theory of 'diminishing returns' is generally used. Extra spending on promotion, environmental management research or new product development in the long run should generate increases in returns at least equal to the expenditures. Further expenditure is justified only to the point when the marginal return ceases effectively to exist. It is important to outline when you plan to spend parts of the environmental management budget. Costs may be phased over different time-scales: weekly, monthly or quarterly, etc.

As part of the environmental management budget, you may find it useful to calculate:

- value added per employee;
- cost per employee;
- cost per product or process.

It is possible to extend this list substantially. Every business has its own unique environmental characteristics, philosophy of management, culture, objectives and methods of working, so it is easy to imagine new 'measures' or other ways of using the environmental management budget.

ISO 14001 quick cost guide

Calculating the costs of introducing ISO 14001 into a company can be summarised as follows.

Direct costs

The direct costs are readily accessible and easy to calculate.

Registration

The registration fee is a signing on fee that basically covers the administration and paperwork issued to a company when joining an accreditation body. Some of these bodies are listed for reference in Appendix 3. The fee is not refundable if the company decides not to progress further or to change the accreditation company.

Assessment fees

Most accreditation companies work on a fee basis. The fees are, approximately, £550 per day; this figure represents an average daily fee for assessment. Half-day fees are available and are slightly higher, about £300. There are three assessment stages to final certification. As a bare minimum, a company may need only 3 days of assessment, which would equate to 5 man–days in total: the first two assessments may be undertaken by one assessor and the final assessment stage will have a team of three for 1 day.

For those with a little bit extra in the petty cash tin, and with the commitment for a speedier implementation process, additional assessment days can be accounted for. One way of speeding up the implementation process and minimising the likelihood of failure is to have an extra assessment day before assessment stages 2 and 3. Preassessment assessments require the presence of only one assessor. The benefit of extra assessment days is that reducing the time required for implementation can save indirect costs. Also, motivation levels tend to stay higher and agendas are less likely to change with a shorter implementation period.

Annual audit costs

To ensure that the standard is maintained, and continually improved, an annual audit is mandatory under ISO 14000 guidelines. For those companies achieving the standard for the first time, audits are required biannually for the first 2 years. These are 1-day audits with one assessor, and costs for the year will add up to £1,100 (2 × £550.00). After the 2-year period, audits are conducted annually, again on the basis of requiring one assessor for 1 day, and thus costing £550.00 for the year.

Indirect costs

The extent of the indirect costs will depend on the level of your enthusiasm and the size of the environmental budget. It will also depend on the size of the working group – the bigger the group, the bigger the overhead allocation – and how often the group meetings are held and how long they last. The list of possible indirect costs likely to be incurred is identified above.

8   Preparing environmental operational procedures

The operational procedures provide the structure of the environmental management system. They are designed to explain and provide the control of processes that ensure that an environmental management system operates effectively. The procedures outlined here are based on the requirements for ISO 14001 and are designed to provide some direction, by way of format, for you to construct your own set. The creation of these procedures will allow you, as environmental manager, to implement change and monitor improvements in the system. Any environmental management system is dynamic in nature, and must be flexible enough to incorporate change. The operational procedures control the internal and external changes to business activities.

The creation of an operational procedure will emerge naturally from the corporate environmental plan. The procedures are, or should be, designed to help you attain your company's targets and objectives. Think of a procedure as a cake recipe. Each recipe is a list of ingredients that, when mixed together and put into the right environment, will produce the desired cake. If, after sampling the cake, you find that it is not to your taste, do not be afraid to change the recipe. To continue the culinary analogy, if you like the cake, keep trying to improve the taste (Figure 8.1). Installing ISO 14001 is not the end of the environmental quest – it is only the beginning. It is a continuous process of environmental and operational improvement.

If, for example, a product specification or its process of manufacture is changed, it is important to check whether the change has an environmental impact. One of the key operational procedures is that of 'general operational control'. This procedure requires the preparation of an engineering change order for any change to the product or production process – an example is given later in this chapter.

Setting the format of the procedure is also important. As you prepare each procedure, you must state its purpose at the beginning as well as naming the person responsible. If there is paperwork to be raised or a register to be completed, identify too what is to be completed and the person responsible for doing this.

Figure 8.1 Creating an operational procedure is like baking a cake.

Eleven operational procedures are used to demonstrate common operational requirements that should apply to most companies, as follows:

- general operational control;
- legislative compliance;
- identification of environmental aspects;
- environmental training;
- environmental record keeping;
- EMS monitoring and measurement;
- internal and external environmental communications;
- emergency preparedness and response;
- hazardous material spillage control;
- waste handling and disposal;
- heating oil storage and bund inspection.

If you can, describe the procedures using the same format (i.e. purpose, scope, responsibility, procedure). The format given below is relatively simple, but clear and uncomplicated in presentation and understanding.

Purpose

You have an obligation to the users of the procedure and to the assessor to state what the procedure is there to do. A clear, one- or two-sentence statement is really all that is required.

Scope

Explain the scope of the procedure. Set out why the procedure has been introduced. Detail what it is going to control, how this will be achieved and the benefits this will bring to the company.

Responsibility

Specify those people responsible for putting the procedure into action. This is quite likely to involve a number of individuals throughout the company. For example, for most if not all procedures, you and or your working group will have overall responsibility for implementation. You may then require input from other people in other departments. There will be requirements for engineers to write and check operational procedures or, perhaps, for someone from purchasing or stores to write and check flow diagrams for goods in and goods out procedures.

Procedure

When writing the procedural part of the process, have in mind three stages to be addressed. The first is the introduction of the procedure. Explain in detail what it is designed to achieve; where it relates to existing operational procedures, it is important to specify which ones.

The second stage is to state clearly the process that will take place when the procedure is actioned. The final stage is for you to explain what the changes may be and whether there is documentation to be raised, and to name the person responsible for completing the paperwork and authorising any changes. It may be useful, when detailing some procedures, to bullet point key duties or activities that need to be performed by the responsible person or persons.

When detailing the procedural element of one operational process, take the opportunity where applicable to refer to other operational processes if they interrelate. As an example, one operational process may explain legislative compliance. To direct the reader to the list of legislation, provide a reference to the environmental manual or to the folder containing the list. Cross-referencing of documents or procedures is a requirement of the standard and is, in any case, a valuable and helpful practice.

The eleven common procedures are now described, accompanied in each case by an example of the completed procedures for your direct use or guidance.

Operational procedure – general operational control

Purpose

This procedure forms part of the environmental management system (EMS). It describes the methods for controlling and monitoring environmental aspects.

Scope

The environmental aspects need to be controlled and monitored in order to develop the positive and reduce the negative impacts. Such changes will ensure continuing environmental improvement.

Responsibility

The environmental manager, supported by the members of the environmental working group, is responsible for the overall control and monitoring of the environmental aspects.

It is the responsibility of the company's engineers to ensure that, if processes or materials change, any such changes are declared on an engineering change order (ECO). The environmental manager is responsible for the review of completed ECO forms.

Procedure

An environmental aspect can change when either the process or the materials involved in the aspect are altered. In all cases where changes occur, the environmental manager is to be advised.

If processes or materials for a particular contract are changed, or modified, the engineer responsible must complete an ECO. The ECO includes a section specifically for stating whether or not there is a positive or negative change in the environmental impacts of the aspects in question.

When completed, the ECO form is to be forwarded to the environmental manager, who will review the changes and determine whether any environmental action is required.

Operational procedure – legislative compliance

Purpose

This procedure ensures compliance with relevant environmental and other legislation.

Scope

Environmental legislative requirements have an impact on the day-to-day running of all companies. They are met and controlled by the EMS as one of the requirements of the EMS.

Responsibility

The environmental manager is responsible for ensuring that the company is aware of, and complies with, all relevant environmental legislation. In addition, the environmental manager is responsible for ensuring that the company is provided with the latest information regarding new or proposed environmental legislation.

Procedure

The environmental manager should maintain links with local and national government environmental departments and agencies in order that the latest information is available. Such links should also include:

- regular visits to relevant web sites;
- regular contact with the local council and the regulatory agency(ies);
- subscription to appropriate environmental newsletters etc.

Any changes in legislation that affect the company should be communicated by the environmental manager to management and employees. The environmental manager should review all such changes and, if necessary, initiate the development of a new or revised procedure and update the Register of Regulations.

Operational procedure – identification of environmental aspects

Purpose

This procedure defines the methods used for the identification of environmental aspects within the company.

Scope

This procedure describes the methods that should be used to identify and monitor existing and new environmental aspects within the company.

Responsibility

The environmental working group (EWG) is responsible for identifying and defining all of the environmental aspects within the company. Individual members of staff are responsible for correctly reporting the introduction of new processes, products or substances relating to or affecting their work or the location in which they work.

Procedure

This procedure involves the identification and reporting of aspects of company operations that have or may have environmental impacts; these include:

- *Office activities:* including the generation of waste and the use of recyclable products such as printer cartridges.
- *Electrical/mechanical design:* including the generation of waste and the consideration of environmental impacts as a design function.
- *Electrical manufacturing:* including the generation of specialist waste and the reuse of wire/cable, etc.
- *Mechanical manufacturing:* including the generation of specialist waste and the reuse of metal/wood, etc.
- *Purchasing and storage:* purchasing covers any form of product that is supplied to the company and includes the use of customer-supplied environmental information. Storage covers the use of hazardous and non-hazardous storage facilities for chemicals, substances, products, manufacturing components and waste.

Details of environmental impacts that are relevant to these areas are recorded in the register of environmental aspects and impacts (see Chapter 9).

New processes, products or substances

For the company EMS to remain effective, it is essential that any new processes, products or substances that are used or adopted within the company are identified and recorded.

The person responsible for introducing a new product, process or substance within the company should complete a new product introduction (NPI) form and forward it to the environmental manager. The environmental manager is responsible for examining all new NPI forms with respect to environmental implications, and for taking the appropriate action (see Chapter 9 for further details).

Criteria for determining significant environmental impact

All environmental aspects or effects are graded from 1 to 5, with 1 meaning low significance and 5 meaning high significance. This grading determines whether or not there are likely to be any significant environmental impacts.

The gradings are determined using the criteria of pollution effect and required resource use. All grading should be recorded in the register of environmental aspects and impacts (see Chapter 9).

Operational procedure – environmental training

Purpose

This procedure defines how company personnel are trained within the EMS.

Scope

This procedure covers the induction and refresher training on environmental matters that is given to all company employees.

Responsibility

The environmental manager is responsible for organising the induction training of all new employees with respect to the EMS and environmental issues, and for the subsequent and periodic refresher training of all employees.

It is the responsibility of managers and supervisors to ensure that personnel for whom they are responsible are adequately trained and are competent to complete a task before carrying out any process or operation. This is particularly important when the process or operation could cause environmental impacts.

Induction training

All new employees should receive a minimum of a 2-hour training brief on the need for environmental awareness in the work situation. The training will include the general topics of waste collection, material reuse, use of manufacturing procedures, etc., together with specific information that is relevant to the employee's type of work.

All new employees should receive general training in the correct actions to be taken in the event of an environmental emergency, such as oil spillage or fire. Employees who will be working in hazardous situations or using potentially hazardous substances should receive specific training.

Refresher training

This type of training is usually conducted after a change in established procedure or before the introduction of a new procedure. The training will be tailored to suit the personnel involved, but will always include reminders on the need for environmental awareness and any changes in emergency preparedness. Some programme of refresher training should take place on a periodic basis, say every 2–3 years, for all staff.

Training syllabuses

The environmental manager, in conjunction with the environmental working group, is responsible for the development of induction and refresher training syllabuses. The syllabuses, once reviewed and proved, should be retained in the environmental documentation folder. Training syllabuses should be reviewed regularly as part of a management review.

Training schedules and records

Details of employees who are to receive training, together with training topics and other details, should be recorded on the company environmental training schedule. Completed training forms should be retained in the environmental documentation folder.

Details of all environmental training that an employee receives should be recorded as part of that employee's training record and placed on file.

Operational procedure – environmental record keeping

Purpose

This procedure defines the method of identification of environmental records, the duration of their storage on file, where they are to be stored and how replacements are updated.

Scope

This procedure applies to all records and forms used as part of the EMS.

Responsibility

The environmental manager is responsible for the storage and maintenance of all records and completed forms that are part of the EMS.

Procedure

All environmental records and forms should be forwarded to the environmental manager, who will store them in the environmental records folder. Each record and form should be dated and retained in the folder for at least 2 years.

The environmental manager should examine all archived environmental records and forms periodically. Those that have been archived for 2 years or longer should be reviewed and a decision made as to whether the document is still live or should be archived.

Environmental records and forms are defined as any records or forms that relate to or contain or incorporate any information regarding the EMS, they include:

- internal audit reports;
- audit schedules;
- procedural changes;
- management reviews;
- records of waste disposal;
- assessor audit comments.

Environmental records and forms should be indexed by a serial number and/or date to ensure that they are traceable. The keeper of the environmental manual should maintain details of forms issued and archived.

Operational procedure – EMS monitoring and measurement

Purpose

This procedure defines the methods utilised for the recording of information that is used to track performance and conformance to the environmental objectives and targets.

Scope

It is essential that the performance of the EMS is monitored in order to assess the need for improvements. This procedure describes the methods used to carry out the monitoring.

Responsibility

The environmental manager is responsible for carrying out the monitoring and recording of information and the subsequent assessment of performance.

Procedure

One of the most effective means of monitoring the EMS is the routine observation of compliance with the procedures set out in this book. The environmental manager should regularly tour the premises and observe procedural compliance.

Ongoing assessment

The environmental manager should collate the receipts from the waste collection agencies and tabulate the quantities listed for each type of waste. These figures should be transferred to a spreadsheet showing the different types of waste quantities per month. The spreadsheet will enable target environmental performance figures to be set and will provide tangible results to be presented to employees.

Operational procedure – internal and external environmental communications

Purpose

This procedure describes the paths for optimum internal and external communication on environmental matters between the environmental management personnel and employees and interested external parties, such as the local authority or environment agency.

Scope

This procedure applies to internal and external communications relating to environmental matters.

Procedure

Company personnel and any external parties who have any opinion, comment, complaint or praise to make regarding environmental issues need to be able to make effective contact. It is therefore essential that the environmental management personnel are made directly aware of such matters in order that a suitable response can be made. It is equally important that information regarding consequential actions is fed back to the party concerned.

Internal communication

Internal communication regarding environmental matters should be directed to the environmental manager or deputy. The person receiving it

should record all such communication in the environmental communications log. If a formal response is deemed necessary, a document folder
should be created to retain all correspondence associated with the matter.
The folder should also contain written details of all related telephone
conversations, e-mails and meetings, etc.

External communication

All communications regarding environmental issues received from an
external source should be directed to the environmental manager or deputy.
Details of the communications should be recorded in the environmental
communications log. If a formal response is deemed necessary, a document
folder should be created to retain all correspondence associated with the
matter. The folder should also contain written details of all associated
telephone conversations.

For the purposes of external communication, the environmental
manager should establish a means of publicising any notifiable environmental incident and any changes or additions to significant environmental
aspects. The definition and relevance of a notifiable environmental incident
and associated publicity guidelines should always be confirmed in writing
with the appropriate agency. A notifiable incident is one that involves the
general public or that will eventually have an impact on the general public.

The following agencies are typical of those that should be informed in
the event of a notifiable environmental incident or of any changes or
additions to significant environmental aspects:

- local authority;
- environment agency;
- water authority;
- local press;
- certification body (e.g. BSI).

All relevant correspondence and communications should be approp-
riately logged and recorded, and a copy should be placed in the
environmental documentation folder.

Operational procedure – emergency preparedness and response

Purpose

This procedure outlines the approved system in case of an environmental
accident and prepares the company for this eventuality.

Scope

This procedure provides guidelines for emergency situations.

Responsibility

In general, workshop maintenance personnel manage an environmental accident response. The senior person present is to be available to assist in decision making and the authorisation of actions.

Procedure

The types of environmental emergency possible include fire, oil leakage and chemical/hazardous material spillage.

Fire

In the event of fire, personnel must follow the fire evacuation procedure and the fire service should be contacted as quickly as possible.

Upon arrival, the fire service should be advised of the presence, type, quantities and location of hazardous/flammable materials within the premises and of any special handling or other precautions to be taken into consideration

Oil leakage

Oil leakage from the heating oil storage tank will be retained by the surrounding bund. If a leak does occur, however, contact the fire service for advice. In addition, if the premises are leased, contact the landlord and the local authority. Next, contact a licensed contaminated water or soil contractor and request a clean-up immediately.

Chemical/hazardous material spillage

Contact workshop maintenance personnel.

When the spill has been contained and cleaned up, and the residue from the clean-up operation is ready for disposal, contact a licensed contractor and request a collection.

Follow-up actions

Any incident that requires the actions of spillage response personnel or external assistance must be documented and reported as a non-conformity. This will assist in identifying the cause of the incident and the introduction of any necessary remedial actions.

Operational procedure – hazardous material spillage control

Purpose

This procedure describes the actions to be taken in the event of a hazardous material spillage.

Scope

This procedure establishes the way in which spillages of hazardous materials should be dealt with.

Responsibility

Workshop maintenance personnel are responsible for controlling and cleaning spillages of hazardous materials.

Procedure

Hazardous materials are those whose spillage could lead to a serious adverse environmental impact. The company premises may house some hazardous materials for cleaning purposes and also for use during some processes, including:

- solvents;
- conditioning and cleaning fluids;
- remover fluid;
- electrolytes;
- tin powder as a colloid.

If a hazardous material spillage occurs, workshop maintenance personnel should be contacted as quickly as possible.

Immediate actions

- Identify the substance involved.
- Warn other personnel.
- If necessary, create an exclusion zone and consider evacuation.
- Wear protective equipment as required.
- Identify the source of the spillage.

Follow-on actions

Use appropriate material to clean up the spillage.

- All soiled or contaminated clean-up material must be stored in a suitably identified bag or container pending collection and disposal.
- If more than one material has been split, try to ensure that the materials do not mix and that, when cleaned up, the materials are stored in separate bags or containers.

Preventive action

All hazardous materials and substances should be stored in the correct storage area in clearly identified containers. Under no circumstances should hazardous materials or substances be stored in unmarked containers. If decanting is necessary, this should be carried out in a designated area. Highly flammable materials and substances should be stored in a separate designated area.

Containment action

Empty containers for hazardous materials and substances should be stored in the designated storage area pending authorised collection and disposal.

Documentation

Any incident that requires attendance by workshop maintenance personnel must be reported and documented as a non-conformity. This will assist in identifying the cause of the spillage and the introduction of any necessary remedial actions.

Operational procedure – waste handling and disposal

Purpose

This procedure defines the methods to be used for the handling, storage and disposal of liquid and solid waste produced in the office, workshop and stores areas of the company.

Scope

The day-to-day running of any company generates waste – paper, plastic, printer cartridges, etc. – much of which can be recycled. This procedure covers the handling, storage and disposal of office-, workshop- and stores-generated waste.

Responsibility

All employees should be trained to dispose of the waste that they produce in a responsible manner. That is to say, they should be responsible for disposing of waste in the waste receptacles and storage areas provided. Individuals should also be responsible for the identification of items such as toner and ink cartridges, wire and metal and wood offcuts that could be recycled.

- The environmental manager is responsible for providing and overseeing the correct means of handling, storage and disposal of waste.
- The environmental manager is responsible for the placing of clearly marked, for example 'plastic' and 'paper', waste collection boxes and containers throughout the company, and for designating waste storage areas.
- The environmental manager is responsible for the nomination of personnel for the emptying of waste collection boxes into suitably marked containers pending disposal and for the identification, handling and storage of all cardboard waste, batteries, electronic components and fluorescent tubes.
- The environmental manager is responsible for the identification and use of suitable waste collection companies. All such companies must be in a position to provide proof that they have all relevant and current licences and certificates.

Categorisation, storage and disposal

- All solid and liquid waste should be categorised, stored and disposed of correctly.
- Any metallic waste, ferrous or non-ferrous, should be identified, categorised and stored correctly in clearly marked containers with the assistance of workshop personnel. Particular care should be exercised when storing oily swarf because of the possibility of spontaneous combustion.
- Any form of clean paper, such as computer printouts, magazines and thin card, should be placed in the boxes labelled 'Paper'.
- All forms of plastic, including cups and packaging materials, should be placed in the boxes labelled 'Plastic'.
- Any dirty or non-recyclable paper, plastic or other form of waste should be disposed of in the rubbish bins labelled 'Rubbish'.
- Waste cardboard boxes should be collapsed to minimum volume and stored in a safe and dry storage area pending disposal.
- The environmental manager should be consulted regarding the suitability of items such as toner or ink cartridges for recycling. If items are not suitable for recycling, advice should be given as to the best method of disposal.
- Fluorescent tubes should be stored in their original packing for safety and clearly labelled. When there are sufficient tubes for disposal, the environmental manager should identify the correct disposal method.

- If possible, liquid waste, including paint, solvents and thinners, should be identified and then stored in suitably marked containers pending collection. Under no circumstances should liquid waste be disposed of down drains or onto the grounds surrounding the plant or buildings whether within or outside the boundary. The environmental manager should be consulted when there is any liquid waste stored for disposal.
- Batteries, drums and any other items that contain or may have contained mercury, lithium, cadmium or other metals or chemicals must be stored separately pending correct disposal.

Full waste boxes should be emptied into suitably marked containers located in the waste storage area. When the containers are due to be emptied, possibly on a specific collection day, the environmental manager should be informed so that the containers can be collected.

The procurement manager must advise the environmental manager when there is a need to dispose of waste, whether cardboard, batteries, metals, oils or hazardous waste.

Waste collection

All categorised liquid and solid waste for recycling and disposal must be collected by licensed and authorised waste collection organisations. Personnel responsible for transferring waste to the collector should check with the environmental manager whether the collector is approved and has provided evidence of holding the correct licences and certificates.

Waste categorised as rubbish should be placed in the council-supplied 'wheelie bin' for collection by the council. Items of rubbish that are too large to fit into the 'wheelie bin' should be stored and the council contacted by the environmental manager or nominated person for advice regarding collection.

The collection of waste should be recorded in the waste collection log. The log records the date, type and weight of waste, and who collected it. The log should be signed by the person collecting the waste as well as by the company's representative. Note that, when the weighing of waste is not possible, quantities should be estimated and recorded as estimated.

Operational procedure – heating oil storage and bund inspection

Purpose

This procedure describes the method for replenishing the heating system oil tank and for inspecting and emptying the oil tank bund.

Scope

When fuel oils are being delivered, there is the chance of a spill taking place and contaminating any rainwater retained in the bund. The aim of this procedure is to minimise the risk of this occurrence.

Responsibility

The procurement manager or nominated deputy is responsible for all aspects of heating oil delivery and replenishment. The environmental manager is responsible for the regular inspection of the bund and for arranging for the removal of any contaminated water.

Tank replenishment

The oil tanker driver must be directed to report to the procurement manager or deputy upon arrival on site. The procurement manager or deputy should supervise all aspects of the replenishment operation, including:

- checking the initial and replenished levels of the tank;
- ensuring that the tanker driver is present during the replenishment operation;
- ensuring that tank valves, filling point, etc. are opened, operated and closed correctly before, during and after the replenishment operation;
- ensuring that any oil spillage is cleaned up promptly and correctly;
- ensuring that the correct paperwork is completed before the tanker's departure.

When the replenishment operation has been completed, the procurement manager should inspect the bund. If any free oil or oil-contaminated water is present, the environmental manager should be informed.

Bund inspection

All inspections of the heating oil storage tank bund should include an examination of the tank, piping and fittings for leaks and damage. The area surrounding the bund should also be inspected to ensure that there is no seepage from the bund. Suitably qualified persons must regularly examine any electrical wiring and components in the vicinity of the bund.

- The environmental manager should arrange for the bund to be inspected at least weekly, and more frequently in the event of heavy rain or snowfall.
- The environmental manager should arrange for contaminated water to be removed by the designated haulier.

Bund leakage

The environmental manager should be informed immediately in the event of leakage from the bund. The sequence of actions taken then will be governed by the severity of the leak.

Slight leakage

If possible, locate and plug the leak. Use any suitable containers to retain the contaminated fluid. Request assistance from your local council and/or clean-up contractor.

Severe leakage

If the contents of the bund leak out, report the matter as quickly as possible to the Environment Agency and your local council. In addition, request assistance from a licensed clean-up contractor.

Summary

As mentioned earlier, the procedures identified in this chapter are designed to represent typical procedures that would be found in most companies and written in a format that would satisfy ISO 14001 procedural guidelines. The numbers of procedures required will vary from company to company, as will their length and complexity. The best procedures are not necessarily long and complex – keep them clear and simple. The procedures you write should support, and enable, the achievement of the objectives and targets in your environmental management plan. Remember also that, as your objectives and targets change, so your procedures will change.

9 Documentation assistant

This chapter is designed to assist with the development of the required documentation for ISO 14001 and to explain the uses of the various documents if you are implementing the standard. The documents discussed below would be kept in the environmental documentation folder (EDF). They provide an indication of the standards requirements and are arranged in no particular order. The format of the documents and the breakdown of sections within them are indicative only and will change to suit your company's activities and reporting style.

Audit cover sheet

The audit cover sheet (Figure 9.1) identifies the location, department and the type of audit to be carried out, e.g. system, procedure or other element. It is the audit team that will undertake the audit, and the names of team members should be inserted together with the name of the person, or group, to be audited. This provides a quick reference as to whether the audit has been completed, and by whom, and where it was conducted.

The audit check sheet

Contained within the audit check sheet (Figure 9.2) are a number of example questions that have been tried and tested, and from which you can select when auditing the various elements of the environmental management system. Obviously, you can use your own audit check sheet where applicable. If you operate any ISO auditing system, the questions that you currently ask can be adapted very easily for environmental auditing purposes.

There are a number of different elements within the ISO 14001 standards that need to be assessed, and it is not feasible or even desirable to ask, say, a dozen questions per element; two questions per element will be sufficient. As you can see from Figure 9.2, even two questions per element amounts to a considerable number to satisfy the audit.

The correct response to each audit question must include the actual

Location	
Department	
Audit type	
Date	

Lead auditor	Name:
Auditor	Name:
Auditor	Name:
Auditor	Name:

Personnel involved

Figure 9.1 Audit cover sheet.

answer to the question as well as the location within the environmental management manual or other folder, log, etc. where the answer can be found. You will find that the full range of questions listed in the audit check sheet may need to be answered by several people within the company. For example, the environmental manager may answer questions relating to environmental policy, while the engineering manager may answer in relation to operational procedures. Of course, there may also be questions that require a combined answer from more than one person, particularly if recommended engineering changes require product design change input.

If an audit is to be valid, it is not sufficient just to give the correct answer to a question. The auditor will require that the person answering the question provides evidence of compliance, i.e. they must be able to identify the document, folder or procedure where the answer to the question can be found. If the answer given is correct, a 'pass' result will be entered; if incorrect, a 'fail' will be entered. A decision will be made on whether the audit has to be readministered or further training provided, depending on the acceptable level of the number of 'fails' per audit.

Audit schedule

The audit schedule (Figure 9.3) provides a record of personnel who are to be audited in accordance with the requirements of the environmental management system. It is another quick reference document to identify those persons within the company who have been audited and those who have not. You can use either of two methods to identify those who have been audited. You could list only those persons who have been audited, or you could list all persons within the company and just date and sign against each name as they are audited. Once the audit has been completed, the auditor responsible will print his or her name and sign the document.

Audit notification

Before beginning the audit, you will need to provide the person to be audited with advance notice. Under the ISO 14001 guidelines, advance notice equates to a minimum of 7 days. This is a useful practice as it gives the auditees time to familiarise themselves with the environmental manual and procedures, and other relevant pieces of paper. This is a simple notice that may take the form of an interoffice memorandum.

To:

Auditee:

Date/time:

Location:

Audit topics:

Auditor:

Under its accreditation to ISO 14001, the company is required to carry out regular audits of the environmental management system (EMS) to ensure that it continues to meet its objectives and that personnel are complying with procedures.

The audit requirements are set out in an environmental management procedure contained in the environmental management manual, and are explained in Chapter 10.

The audit notification form provides advance notification of the following:

- who will be audited;
- when and where the audit will take place;
- details of the topics to be covered by the audit;
- the name of the auditor or auditors.

At the conclusion of the audit, a report will be prepared, by the auditor, for circulation to auditees and members of senior management who are involved in the next management review. Any non-conformity, or the need

No.	EMS requirement	Evidence of compliance	Result
1	GENERAL REQUIREMENTS Establish that a documented environmental management system exists within the company		
2	GENERAL REQUIREMENTS Establish that the environmental management system has been implemented.		
3	ENVIRONMENTAL POLICY Establish that a corporate environmental policy exists.		
4	ENVIRONMENTAL POLICY Establish how the environmental policy has been made publicly available and displayed within the company.		
5	ENVIRONMENTAL ASPECTS Establish how information on environmental aspects is kept up to date.		
6	LEGAL AND OTHER REQUIREMENTS Establish how the list of relevant legislation and other requirements is kept up to date.		
7	OBJECTIVES AND TARGETS Establish how significant aspects are considered in the setting and reviewing of environmental objectives and targets		
8	EVIRONMENTAL MANAGEMENT PROGRAMME Establish that new products, projects, services or activities are covered and included within the programme.		
9	STRUCTURE AND RESPONSIBILITY Verify that top management has a nominated representative who has total authority for environmental management.		
10	TRAINING, AWARENESS AND COMPETENCE Establish that the company has identified its environmental training needs.		
11	COMMUNICATION Verify the procedure used for internal communications within the company.		
12	ENVIRONMENTAL MANAGEMENT SYSTEM MANUAL Establish the existence of a documented EMS system in the form of a manual.		

No.	EMS requirement	Evidence of compliance	Result
13	DOCUMENT CONTROL Establish the existence of a working document control procedure.		
14	OPERATIONAL CONTROL Verify that procedures state the required operating criteria.		
15	EMERGENCY PREPAREDNESS AND RESPONSE Verify that procedures are in place that identify the potential for, and the response to, accident and emergency situations.		
16	EMERGENCY PREPAREDNESS AND RESPONSE Verify that the company tests such procedures periodically where practicable.		
17	MONITORING AND MEASUREMENT Verify that there is a documented procedure for periodic assessment of compliance with all relevant legislative and other requirements.		
18	NON-CONFORMANCE AND CORRECTIVE ACTION Verify that there is a documented procedure that defines the location of responsibility and authority for handling system non-conformance and corrective actions.		
19	RECORDS Establish the existence of procedures covering the maintenance and availability of environmental records.		
20	ENVIRONMENTAL MANAGEMENT SYSTEM AUDIT Determine that environmental management systems are audited regularly.		
21	MANAGEMENT REVIEW Establish that the company carries out management reviews.		
22	MANAGEMENT REVIEW Verify that the management reviews are documented.		

Figure 9.2 Audit check sheet.

Auditee	Audit date	Auditor	Auditor signed

Figure 9.3 Audit schedule.

for a follow-up audit arising from this audit, will be processed in accordance with the non-conformity report (NCR) procedure.

Audit report sheet

Following any audit, a report sheet (Figure 9.4) needs to be completed by the auditor giving an account of the audit and all outcomes. The report sheet has a results summary section (see Figure 9.4) comprising a series of one-line statements of outcomes. There are four possible outcomes that can be recorded in this section:

- Fully compliant – the audit has been completed satisfactorily and no further action is necessary.
- Recommendation – the audit has been completed satisfactorily, although a recommendation has been made that, when effected, will improve current working practice.
- Observation – the audit has been completed satisfactorily, although an observation has been made that, if not effected, will result in a non-conformity.
- Non-conformity – the audit has not been completed successfully and a non-conformity has resulted.

Audit activity schedule

The audit activity schedule (Figure 9.5) is a calendar representation of all activities included in the environmental management system, including the system itself. An annual audit is required for each of the activities identified in the schedule. What the assessor wants to see is an annual schedule, identified by month, that indicates when each activity audit is to be carried out.

Identifying which day the audits are to be carried out will be through

Location	
Audit ref.	
Audit type	
Date	
No. of pages	

Senior manager/Director	Name	Signature
Lead auditor	Name	Signature
Auditor	Name	Signature
Auditor	Name	Signature
Auditor	Name	Signature

Audit results summary
Fully compliant
Recommendation
Observation
Non-conformance

Figure 9.4 Audit report sheet.

judgement, and such dates will slip. You may find that two or three audits are being carried out each month because of slippage. This is acceptable and will probably happen. All audits specified in the schedule, however, will need to be completed before any assessor's audit or you will incur a non-conformity.

Duty of care – waste transfer note

As previously mentioned, you – as environmental manager – are responsible for the disposal of company waste. Before the waste is passed on to a third party, ensure that they are licensed to dispose of that type of waste and that the form shown in Figure 9.6 is completed before the waste is handed over. The duty of care form, or the waste transfer note as it is sometimes called, is the company's receipt to say that every precaution has been taken to ensure that the third party will dispose of the waste in an environmentally friendly manner.

The duty of care form is signed by you, the transferor of the waste, and the third party, the transferee who is disposing of the waste on your behalf.

The form is divided into four main sections. The first, section A, requires

Workshop/production department

	Year	2001																								
	Month	January				February				March				April				May					June			
	Week ending	10	17	24	31	07	14	21	28	07	14	21	28	04	11	18	25	02	09	16	23	30	06	13	20	27
Activity to be audited	Week no.	1	2	3	4	5	6	7	8	9	10	11	12	13	14	15	16	17	18	19	20	21	22	23	24	25
Waste handling, storage and disposal			▨																							
Hazardous materials spillage control								▨																		
Emergency preparedness													▨													
Heating oil storage and bund inspection																	▨									
Reuse of wire and cable, wood and metal																				▨						
Discharge of liquid waste																									▨	

Figure 9.5 Audit activity schedule.

you to specify the type of waste being disposed of, how the waste is contained, for example in sacks or drums, and the approximate weight of the material. This part of the form also provides you with the waste recording information that is entered into the waste recycling log (see Chapter 10).

Enter the company details, name and address in section B together with your credentials for disposing of the waste. Of the four categories listed on the form, it is likely that the 'waste producer' category will be the most applicable unless you are a local authority or waste recycling company.

Section C is to be completed with the details of the company disposing of the waste on your behalf, the transferee. Again, specify what kind of waste disposal company you are dealing with. If it's a private company, enter the details together with its licence number. If it is a local authority disposing of your waste, a licence number is not required. If you are using a registered waste carrier, insert the carrier details and licence number. If you are using another option, insert the details and give an explanation for this choice.

In section D, you are required to provide details of the collection point for the waste and the date of collection. If there are multiple waste pick-ups, provide details of a waste collection period, giving time (2–5 p.m.) or day (Tuesday–Thursday) ranges. The form is then signed by both parties to confirm that the waste has been passed responsibly to another party for safe disposal. Give a copy of the form to the transferee and place your company's copy in the environmental documentation folder.

Environmental plan targets and objectives

The details of your company's environmental targets and objectives may be given in a table (Table 9.1). They are not listed in any particular order; Table 9.1 acts merely as a reference point. Taking the first objective in the table, the detail of the objective is clearly stated. A system will be established, and maintained, by the production manager, or other nominated responsible person, to track and record the recycling of paper and cardboard, glass, tyres, batteries, vehicles, electronic equipment or other. If you need to remind yourself how the objectives were created, see Chapter 5 for more detail.

So that some measure of an objective is realised, a target of a 5 per cent reduction in paper and cardboard waste has been set. Instead of the office waste paper and cardboard going into the rubbish bin as usual, it is collected and separated for recycling. You may think that it is possible to improve on a 5 per cent target, but it is useful to begin at a conservative level. It may be prudent in some circumstances, areas where you are unsure of the level of reduction, to use the data from the first year to establish a benchmark figure and not to try to make a measured reduction.

After setting the target, state the method that will be used to achieve it.

Section A – Description of waste

1. Describe the waste being transferred:

2. How is the waste contained?

   Loose ☐   Sacks ☐   Skip ☐   Drum ☐   Other:

3. Quantity of the waste (number of sacks, weight, etc.):

Section B – Current holder of waste (transferor)

1. Full name (print):

2. Name and address of company:

3. Which of the following are you? (Tick one or more boxes)

   Waste producer ☐   Holder of waste disposal ☐   Licence no.:
                         or waste management licence   Issued by:

   Waste importer ☐   Exempt from requirement ☐   Give reason:
                         to have a waste disposal
                         or waste management licence

   Waste collection ☐   Registered waste carrier ☐   Registration no.:
   authority   Issued by:

   Waste disposal ☐   Exempt from requirement ☐   Give reason:
   authority   to register

Section C – Person collecting the waste (transferee)

1. Full name (print):

2. Name and address of company:

3. Which of the following are you? (Tick one or more boxes)

                         Authorised for transport ☐   Specify which of
                         purposes   those purposes:

   Waste collection ☐   Holder of waste disposal ☐   Licence no.:
   authority   or waste management licence   Issued by:

   Waste disposal ☐   Exempt from need to have ☐   Give reason:
   authority   a waste management licence

                         Registered waste carrier ☐   Registration
                                                          no.:

                         Exempt from requirement ☐   Issued by:
                         to register   Give reason:

Figure 9.6 Waste transfer note (duty of care).

Section D

1. Address of place of transfer or collection point:

2. Date of transfer

3. Time(s) of transfer (for multiple consignments, give 'between' dates):

4. Name and address of broker who arranged this waste transfer (if applicable):

Transferor	**Transferee**
5. Signed	Signed
Full name (print):	Full name (print):
Representing:	Representing:

Figure 9.6 Waste transfer note (duty of care) (continued).

Using the conservative approach mentioned above, gradually increasing the amount of paper and cardboard to be recycled is sufficient comment. The person or persons responsible for achievement of the objective should be identified; this may be one person or many persons. It is likely that the environmental manager will be responsible for achieving most, if not all, of the objectives, but some objectives may be the responsibility of the whole company.

It is also important to detail achievement deadlines, or goals, for each objective. At the beginning, it would be reasonable to set annual goals for objective and target achievement. Allow space in the table for recording the status of the objective; most will be ongoing, but some may be achieved, or even abandoned, during the course of a year and before the management review. Finally, put in the start date of the objective; how else can you know when each objective has been achieved and how long it has been in existence?

Environmental management review – agenda

When the new environmental management system has been operating for a year, it is time for senior managers to sit down and review it and the process, and determine what objectives and targets have been achieved. Setting an agenda and the minuting of this review are an ISO 14001 requirement, and the agenda format in Figure 9.7 reflects all those elements of the environmental management system that need to be reviewed, actioned and minuted.

The layout of the agenda should be as set out in Figure 9.7 together with the minuted responses to each topic. The date, time and location of the meeting together with the names of those members of the senior

Table 9.1 Objectives and targets

Item	Objective	Target	Method	Responsibility	Goal	Status	Date
01	Tracking and recording recyclable paper and cardboard	Following the establishment of a base figure by May 00, a minimum of 5% reduction in recyclable paper and cardboard will be set	Gradually increasing the amount of paper and cardboard that is recycled	EM and procurement manager	Feb 01	Ongoing	Feb 00
02	Reduction in office and workshop paper and cardboard waste output	Following the establishment of a base figure by May 00, a minimum of 5% reduction in paper and cardboard waste will be set	Use of memorandums and newsletters to advise personnel on the need to reduce paper and cardboard usage	EM and all personnel	Feb 01	Ongoing	Feb 00
03	Reduction in office and workshop plastic waste output	Following the establishment of a base figure by May 00, a minimum of 5% reduction in office and workshop plastic waste disposal will be set	Use of memorandums and newsletters to advise personnel on the need to reduce plastic waste and the best means of so doing	EM and all personnel	Feb 01	Ongoing	Feb 00

04	Reduction in workshop wooden and metallic waste output	To establish a means of quantifying reductions in workshop wood and metal waste	Use of memorandums and newsletters to advise personnel on the need to reduce wood and metal waste	EM and all workshop personnel	Feb 01	Ongoing	Feb 00
05	Reduction in energy consumption	Reduce energy consumption by a minimum of 5%. To include electricity and heating fuel. Baseline to be determined by May 00	Personnel encouraged to make savings where possible. Lower thermostat settings and more efficient use of heaters. Turn off lights, VDUs, etc.	EM and all personnel	Feb 01	Ongoing	Feb 00
06	Reduction of hazardous material and substances	To establish a means of quantifying reductions in hazardous material and substances output	Use of memorandums and newsletters to advise personnel on the need to reduce the use of hazardous material and substances in products and processes	EM and all personnel	Feb 01	Ongoing	Feb 00
07	Reduction in water consumption	To establish a means of quantifying a reduction in water consumption	Use of memorandums and newsletters to advise personnel on the need to reduce water consumption	EM and all personnel	Feb 01	Ongoing	Feb 00

Note
EM, environmental manager.

Date/time:

Location:

Present:

Item	Topic	Action
01	Review of previous minutes	
02	Report on progress of monthly audits	
03	Those NCRs raised that have been closed out with CARs. The paint-spraying incident was minor and isolated and it was felt that an issued memo would be sufficient.	
04	The use and effectiveness of EMPs was discussed. No problems or difficulties arose.	
05	The use and effectiveness of EOPs was discussed. No problems or difficulties arose.	
06	Personnel responsibilities were discussed and are considered adequate. No points arose.	
07	Documentation and record-keeping were discussed. Current standards are considered adequate, and no points arose.	
08	The need for familiarisation and training was discussed. Current levels are considered adequate. No points arose.	
09	The existing training syllabuses were considered to be current and adequate still.	
10	No critical non-conformities were identified. All CARs have been closed out.	
11	No employee or EWG recommendations or suggestions have been received.	
12	No additions or deletions of objectives or targets – CEP remains unchanged.	

Authenticated: _____ Senior manager/Director

Minutes agreed: _____ Managing Director

Figure 9.7 Environmental management review agenda.

management team and the environmental working group present must be detailed at the top of the form.

The same form can be used, if you prefer, for both setting the agenda and recording the minutes. The example given in Figure 9.7 combines the agenda topic and the minuted responses. The action column provides an opportunity to identify action to be taken and the person responsible for that action. You will notice that the topic examples provided cover the main areas of the environmental management system from audits, procedures, document control, NCRs/CARs and the corporate environmental plan (CEP). You may find that some or all of these have been working satisfactorily over the review period (1 year), but they must still be included in the review agenda.

Review each topic in turn, and comment on where there are changes and where there are no changes for each topic. Some minor changes may have been actioned already. A note should be made of these even if no further action is required.

Once the review meeting has been concluded, the minutes should be finalised and signed off by the senior manager in attendance and the managing director. The managing director needs to sign off the minutes whether in attendance or not to demonstrate top management commitment to the environmental management system.

New process introduction form

You should be well aware by now that, if you are introducing an environmental management system into the company, you are responsible for the operational processes within the company and the products and services leaving the company. Similarly, you are also responsible for the environmental issues that may arise from goods or services coming into the company.

The new process introduction (NPI) form (Figure 9.8) has to be completed for any new product or process coming into the company that may have an environmental impact. All products coming into the company must be assessed for their environmental impact. If they are hazardous in any way, the environmental manager must complete the NPI form. The materials safety data sheet (MSDS) should be attached to the NPI form for instructions on how to store or handle the product. This is not as complicated as it first appears. If the company already deals with any hazardous substances, they will be listed in a register or log as part of the health and safety or control of substances hazardous to health (COSHH) requirements. Any new hazardous products that come into the company will not be in the existing register. For these, simply complete the NPI form, attach the MSDS and record the new product in the hazardous substances register or log.

1. Name of process, product or substance introduced:
    ....

2. Manufacturer – name, address and contact details:
    ....
    ....

3. Hazardous or non-hazardous?
    ....

4. Reason for introduction: ....

5. Date introduced: ....

6. Introduced by (print name): ....

7. Contract details: ....

8. Materials safety data sheet available? ....

9. Are there any environmental implications resulting from the introduction of this
    new process, product or substance? If yes, provide details, together with
    storage requirements and whether any precautions are required, etc.:
    ....
    ....
    ....

Actions:	Yes	No
Data sheet used for COSHH assessment?	☐	☐
Any special storage requirements?	☐	☐
Any special disposal requirements?	☐	☐
Any special clean-up requirements?	☐	☐
Can the product, substance, etc. safely be recycled?	☐	☐
Register of environmental aspects and impacts annotated?	☐	☐

Attach additional sheets if necessary

Figure 9.8 New process introduction form (NPI).

The environmental manager is the person responsible for completing
and signing off the NPI form. When completing the form, the environmental manager may propose a number of different actions. If the new
product or process being introduced is non-hazardous, the form is

completed with the product and supplier names and all of the action boxes are marked 'no' (Figure 9.8). If, however, the new product or process is considered to be hazardous, and this will be determined by reading the MSDS, different actions such as special storage or disposal requirements may be required. Incidentally, if there is no MSDS, telephone the supplier of the product to get one before making any decisions; the supplier is required by law to produce it – the Chemicals (Hazardous Information and Packaging for Supply) Regulations 1994 apply.

If the product is hazardous and you do have an MSDS, tick the 'yes' action box. From the data sheet, determine whether the product needs any special storage requirements. It may, for example, need to be kept in cool storage or in a steel cabinet; again the data sheet will advise you.

The MSDS is comprehensive in providing details of the product supplied. Divided into approximately twenty sections, the MSDS provides a variety of information about the product, including physical and chemical properties, ingredients, first aid and fire-fighting measures, disposal information, toxicology information and ecological information.

The MSDS will also advise whether the product is recyclable or has any special disposal or clean-up requirements. If the answer to any or all of these is 'yes', tick the relevant box. Finally, you will need to determine whether this new product has a significant or potentially significant impact upon the environment. If so, it may have to be recorded on the register of aspects and impacts, and its environmental impact assessed along with those of all the other materials, etc. used by the company.

Supplier questionnaire

Supply chain issues need to be considered and addressed as part of the ISO 14001 standard. Appendix 1 offers a wider discussion of environmental supply chain issues as they relate to larger organisations and SMEs. The discussion here relates to assessing your suppliers' environmental credentials. Unless specified by you, suppliers do not need to have an environmental management system if they wish to supply your company with products or services. You do, however, need to determine the environmental issues as they relate to your company's environmental management system; to this end, you need to have information about all products or service received.

The supplier questionnaire (Figure 9.9) asks your company's supplier questions about the product that, when answered, allow you to make an informed decision as to whether the new product brings with it environmental risk that is or is not acceptable. If the relevant information is not forthcoming from the supplier, then perhaps another supplier should be considered. You are then, in effect, taking all reasonable precautions to ensure that all business operations are environmentally sound.

Our company is environmentally aware and operates in accordance with ISO 14001. It has BSI accreditation for compliance with this standard.

In order to provide information for our records, it would be appreciated if the questions set out below could be answered. The questions relate to any possible environmental impacts that your products could have.

To: ....

Company: ....

From: ....

Date: ....

Company address

....

....

Telephone number: ....

Fax: ....

E-mail: ....

		Yes	No
1.	Does your company provide material safety data sheets with its products?	☐	☐
2.	Do any of your products produce harmful emissions?	☐	☐
3.	When using your products, are there any special precautions that need to be taken?	☐	☐
4.	Are there any special storage requirements?	☐	☐
5.	Are there any special clean-up requirements?	☐	☐
6.	Can your products be recycled safely?	☐	☐
7.	Are there any special disposal instructions for your products.	☐	☐

Please attach any further information to this form

Thank you for your help in providing this information. It is very much appreciated.

Figure 9.9 Supplier questionnaire.

Environmental training schedule

At some point during the introduction of the environmental system, some element of environmental training will have been given to your company's employees. The training may take the form of either individual or group training sessions, and may be specific in some instances and general in others. For example, specific training may take the form of familiarising a dedicated person with the procedure for heating oil delivery and storage. General training may simply be advising the individual where the environmental manuals and procedures are located for occasional reference needs.

Whatever type of environmental training is administered, it is a requirement of the ISO 14001 standard that a record is maintained. If there is an existing quality training schedule, for example, adopt it or combine all of the company's training needs into the one schedule. Failing that, the schedule given in Figure 9.10 provides an insight into the type of information that it is necessary to record.

Enter the name of the individual who has attended the training course (the attendee) together with the type of environmental training undertaken (topic). Enter the name of the trainer and make any key comments that came from the training. For example, further or additional training may be required or the training may need to be repeated.

Document transmittal advice

The document transmittal advice (Figure 9.11) is central to the document control procedure. Each recommendation is given a unique reference number that is recorded on the revision record inside each document (see Figure 9.12). For example, if the environmental manual is issued to an individual or group, inside the manual on the revision record each intended recipient should be listed and against each name the recipient's transmittal advice reference number should be specified.

The document transmittal advice (Figure 9.11) is made up of two parts. Put the advice number in part 1, again sequential but adding month and year for quick reference, e.g. 001/May/01. The name of the person issuing the document should be entered, together with the date of issue.

Under the heading 'Document Reference', allocate the control number to the document being issued. It is unlikely that there would be more than

Location:			
Month/year:	Department:		
Attendee	Topic	Trainer	Comments

Schedule approved: _____

Figure 9.10 Environmental training schedule.

Part 1 of advice no.: 001/May/01

To:

From:

Date:

Document reference

Document no.	Title	Issue	Revision no.
01	Environmental management manual	Jun 01	00
02	Environmental management manual	Jun 01	00
03	Environmental management manual	Jun 01	00

Action details

☐ New complete document issued to you for retention as registered holder

☐ Complete replacement for an existing document that you hold

☐ Revision(s) to an existing document that you hold

☐ Return removed page(s) to:

☐ Dispose of removed page(s)

☐ Other

Instructions

....
....
....

Part 2 of advice no.: 001/Apr/01

Acknowledgement

When actions are complete, sign and return this section to:

Certified that actions specified in part 1 of this form have been carried out correctly.

Signed: _____   Date: _____

Figure 9.11 Document transmittal advice.

three controlled documents: (1) the environmental management manual (EMM), (2) the procedures and work instruction folder and (3) the document control folder. The issue numbers of these documents are uniform in usage and begin with 01. The number that is issued will depend on the number of locations, or key people, within the company that need copies. If not for the preservation of the rainforests but for ease of administration, try to keep the numbers of documents issued to a minimum.

Using the environmental management manual as an example, the original, or master copy, will be numbered 01. The master copy of all other controlled documents will also be 01. If you consider that copies of the EMM should also go to the production department and/or the administration department, these copies may be numbered 02 and 03 respectively. All controlled documents issued to these departments would also use the same numbers. You are, in effect, allocating a code number to the department to which the controlled documents are being sent.

In the example given in Figure 9.11, under the heading 'Document Reference', three EMMs are being issued – the original, 01, plus two other copies, 02 and 03. The issue date is also entered together with the revision (Rev) number 00. The revision number will obviously change every time the EMM is reissued.

The 'action details' of the advice inform the recipient of the type of change that has occurred. If the system is new, the first documents issued will be new, as will the registered holder of the document. If this is the case, the issuer of the document will indicate this in the first check box.

In the document transmittal advice shown in Figure 9.11, the document control numbers are detailed together with the department and the named person who is to be the registered keeper of the document. The advice will indicate the action(s) that need to be taken. There may be a new manual or a page may need to be replaced by a revised one. There may also come a time when the whole document needs to be reissued. This will occur when the number of changes made within a document become, in your opinion, unwieldy or untidy and the whole document needs to be reissued.

If revisions are made to the document, the appropriate pages, i.e. those that have been changed, will be issued to all registered holders of the document. If this situation occurs, the issuing person will provide further instructions as to whether the pages being replaced (the previous ones) are to be returned. Remember when dealing with controlled documents that all changes have to be recorded together with the actions taken.

When all of the actions requested have been effected, part 2 is completed to confirm this. Part 2 of the document transmittal advice requires the registered keeper of the manual to enter the advice note number as in part 1. The acknowledgement is signed, dated, detached and returned to the issuer of the document. In effect, part 2 acts as a receipt for the information received and the actions taken.

Revision record

Once document changes have been sent to the named recipients, it is those recipients who are responsible for recording the detail of the changes on the revision record – kept at the front of the document. A record must be made of that part of the document affected by the changes; if the whole document is changed, simply state 'all'. Record the issue number of the document, the revision number, brief details of the changes made and ensure that it is dated and signed. Figure 9.12 gives an example.

Induction training syllabus

For every new employee entering the company, and this will include contractors that are with the company for a minimum period of 3 months, environmental induction training is required. The document shown in Figure 9.13 contains details of the induction training syllabus that would typically be used to provide essential environmental information for all personnel joining the company. The training should be conducted by a member of the EWG or by a suitable person authorised by the environmental manager or deputy.

The training covered by this syllabus is designed to be flexible and to be adjusted to suit changes in circumstances, procedures or actions. All employees should receive environmental training in line with the stated

INSTRUCTIONS

This page contains a record of revisions to the contents of this publication. If this page is not replaced by a revision, details of the revision must be recorded in this record.

Part affected:	insert details of the chapter, or section, or section title or page, etc.
Issue:	insert the issue status or date of the revised pages.
Revision no.	insert the revision number of the revised pages.

It is the responsibility of the person having charge of this publication to ensure that this record is correctly and accurately maintained.

Part affected	Issue	Revision no.	Brief details of revision	Date revision inserted	Inserted by
All	Jun 02	00	Initial issue	N/A	

Figure 9.12 Document revision record.

General training coverage

Corporate environmental policy .... ❑

ISO 14001 and the need for environmental awareness .... ❑

Environmental documentation .... ❑

Paper and plastic waste disposal .... ❑

Process waste disposal .... ❑

Recycling of waste .... ❑

Reuse of material .... ❑

Reduction of electricity/water use .... ❑

Office area

Location of toilets and the need for cleanliness .... ❑

Location and operation of drinks machine .... ❑

Location of paper and plastic waste collection points .... ❑

COSHH and substance information book .... ❑

Heat and energy conservation .... ❑

Workshop area

Location of toilets and the need for cleanliness .... ❑

Location and operation of drinks machine .... ❑

Location of paper and plastic waste collection points .... ❑

Location of process waste collection points .... ❑

Location and use of basic personal protective equipment (PPE) .... ❑

Location of hazardous substances in storage and use .... ❑

Emergency actions in the event of a hazardous substance spill .... ❑

COSHH and substance information book .... ❑

External area

Location of waste storage facilities .... ❑

Location of heating oil tank and bund .... ❑

Location of drainage and ventilation .... ❑

Location of other discharge points .... ❑

The associated check boxes should be ticked to confirm that the individual has completed training in each of the topics.

Figure 9.13 Induction training syllabus.

syllabus, including visits to the working areas of the company and a brief tour of the outside area of the business premises.

The training syllabus should cover the main areas of the business, particularly those areas that are the focus for achieving the environmental objectives and targets. Split into four main areas, the training syllabus shown in Figure 9.13 would be relevant for many companies.

The 'general' category is the most obvious place to begin as it requires that each new employee is made aware of the company's corporate environmental policy, i.e. what the company's objectives are and where they are displayed. The new employee also needs to be made aware of the type of environmental management system in operation, and, therefore, the need to be aware of the environment in the course of daily duties. This will require the new employee to become familiar with relevant environmental system documentation and its location within the company.

Practical considerations of the EMS require the employee to be familiar with the correct procedures for the disposal and recycling of paper, plastics, wastes, etc. This goes hand in hand with awareness of energy efficiency, material reuse programmes, etc. that are currently in operation.

Depending on the location in which the new employee works, for example whether he or she is based in the office or in a workshop area, different training procedures will apply, although the same outcome will be realised. For example, waste plastic and paper containers tend to be located in different areas of the company, and emergency actions required in the event of a hazardous substance spillage may be more relevant to the workshop area.

As part of the environmental training programme, it would be prudent for some new employees to be made aware of the external surroundings of the company, particularly those who will be responsible for the maintenance of company heating, draining and ventilation systems, etc. It was recommended earlier that all employees are given an external tour of the premises; the identification of the location of drains and 'rodding eyes' would be relevant only to personnel responsible for company maintenance.

The relevant check boxes in Figure 9.13 are ticked to confirm that an individual has completed training in each of the topics.

Refresher training syllabus

Other that being shorter in duration – it is a refresher after all – the refresher training syllabus is designed to provide periodic, most commonly annual, updates of staff environmental knowledge and awareness of the existing system. It would save time and effort to use a condensed version of the induction training syllabus; how condensed is up to you and your company's requirements. It is likely that some rather than all of the staff will need refresher training depending on their level of involvement with the EMS and the potential for environmental impact from their particular role and department in the company's operation.

The system may specify the degree of refresher training required, i.e. a few hours or 1 day, and the frequency with which it is should be given, e.g. every 6 months or annually. Similarly, an increase in the number of NCRs and CARs being raised may also suggest that additional training is required.

Non-conformity report

The non-conformity report (NCR) document (Figure 9.14) identifies those parts of the environmental management system that are not working. The NCR can be raised by anyone in the company who identifies that a part of the company operation does not conform to EMS procedures.

The NCR is made up of two parts. The first part, the 'originator actions', provides details of the non-conformity and the person identifying the non-conformity. The originator will start by identifying the type of non-conformity. On the form, the originator will specify whether the non-conformity has occurred through a failing of a procedure, the system, or a product or process (see Figure 6.1 for more detail). The originator will identify whether the non-conformity is critical or non-critical.

If the non-conformity is identified as critical, for example a hazardous substance spillage, a verbal as well as written response may be made to the environmental manager for immediate action. If the non-conformity is non-critical, i.e. a procedure that is not being followed as specified, then the appropriate box is ticked and a recommendation is made under item 2 of the form as to the corrective, or preventive, action to be taken. The originator then signs and dates the form and forwards it to the responsible person.

Depending on how the company is structured and the type of system in place, the originator of an NCR may seek the assistance of a senior engineer, manager or auditor for guidance in completing the form. It is likely that the person providing assistance will be the designated person required to complete the second part of the form.

The designated person, senior engineer or manager will complete the second part of the form, detailing the action to be taken to rectify the non-conformity. Following discussion with the originator of the NCR, the actual action(s) agreed will be detailed in item 5. If a corrective action request (CAR) is to be raised, the 'yes' box is ticked and the serial number of the CAR entered on the form. The numbering sequence for NCRs and CARs is consecutive, and quick reference to the documentation folder will tell you the next number to be used in each case.

If a CAR is not to be raised, the 'no' box is ticked and the explanation given in item 5 will state the reason for not raising the CAR. Although a CAR may not be raised, reference to the cause of the NCR in the first place

Originator actions

NCR serial no.:	Location:
1. Type of non-conformity Tick appropriate box: Procedural ❑ System ❑ Process/material ❑ Description, including any EMS requirement, document type/reference (as applicable): Recommended category: Critical ❑ Non-critical ❑	
2. Corrective/preventive action(s) recommended or required	
3. Originator signature/name:	4. Date:

Designated person actions

5. Actual corrective/preventive action agreed	
6. CAR raised Yes ❑ No ❑ CAR serial no.:	
7. Verification – signature:	8. Date:

Figure 9.14 Non-conformity report (NCR).

can be made in the minutes of the management review process (see Chapter 6).

Corrective action report (CAR)

Much has been written about corrective action reports (CARs), particularly in Chapter 7 of this book. The sample form in Figure 9.15 shows what a typical CAR would look like. The completion of this form is likely to be undertaken by an auditor or similarly responsible person within the company.

The form should be identified at the top with the unique CAR serial number (select the next one in the sequence being raised), as well as the NCR number that it is addressing. Then enter the site or location that it refers to – this will be the location where the NCR occurred. State why the CAR has occurred, e.g. whether it was because the NCR resulted from an internal audit or from an observation made by an employee.

Next, enter the name of the originator of the NCR and the date of the NCR. Add the name of the person with whom the NCR was discussed and the part of the EMS that it affects. Detail what needs to be done to meet the requirements of the EMS and then identify where the EMS deficiency arose. Both the auditee (if applicable) and the auditor (if applicable) sign the form. Remember, an NCR may arise as part of an audit or from an employee observation.

The first part of the CAR relates essentially to details of the NCR raised initially. The next part, from point 11 onwards, is about any corrective or preventive action proposed to be specified. This may take the form, for instance, of rewriting part of a procedure that does not work or of clearing up a chemical or oil spillage.

Then state the date for the actions identified in the CAR to begin, and the person responsible for carrying them out. Ensure that both the originator of the CAR and the person responsible for its execution sign the form by way of agreement and understanding. Finally, when all the CAR actions have been completed, the action details are entered and the form is signed off and dated by the person effecting the CAR actions.

Register of aspects and impacts

The process for identifying the environmental aspects and impacts that affect a company are identified in Chapter 4. The register of environmental aspects and impacts, detailed in Figure 9.16, is the place to record them. The aspects and impacts of the company cannot be cast in concrete; the operations of all companies, and the laws affecting them, change with time. They can be added to at any time, particularly if new processes or products that may have a significant environmental effect are introduced. The new process introduction (NPI) form is one tool that the environmental manager can use to identify new aspects and impacts. Aspects may be deleted from the register if the particular product, process or service has been removed from the company's business activities.

The ISO 14001 standard requires that the register records information similar to that detailed in Figure 9.16. Identify the activity, product or service that provides the basis for the aspect. Detail the environmental aspect that has the potential to occur (see Chapter 4 for more detail). Then detail the type of impact that would arise if the aspect occurred. If the aspect has already occurred, and this is unlikely if your system is new, record the numbers of the NCR and CAR that refer to the incident in the final column.

CAR serial no.: 001	NCR serial no.: 001
1. Location Production department	2. Source: audit/observation Audit
3. NCR originator John Smith	4. Date of findings/report 01/01/01
5. Discussed with Engineering manager	6. EMS document section EMM EOP XX
7. EMS requirements Spillage clean-up – contaminated absorbent material not disposed of	
8. EMS deficiency None	
9. Auditee's signature (if applicable) 10. Auditor's signature (if applicable)	
11. Corrective action required Contaminated material to be collected and bagged by maintenance	
12. Preventive action required None	
13. Planned start/completion dates As soon as possible	14. Responsibility for action Production manager
15. Agreed by addressee	16. Auditor's signature
17. Corrective/preventive action taken Memo of clean-up requirements given to maintenance manager	
18. Addressee's signature	19. Date completed

Figure 9.15 Corrective action request (CAR).

Item	Activity/product/service	Environmental aspect	Environmental impact	NCR/ CAR details
01	Use of tinning powder	Potential for spillage	Possible atmospheric contamination	
02	Use of tinning powder	Possibility of accidental spillage	Possible soil and water contamination	
03	Use of copper plate	Direct flow to drains	Water contamination	
04	Use of Freon solution	Direct flow to drains	Environmentally approved washing liquid	
05	Generation of waste paper	Recycle waste paper	Conservation of natural resources	
06	Preparation of wire and cable assemblies	Reuse of wire 'tails' and offcuts	Reduce product costs	
07	Servicing of company cars	Potential for oil spillage	Possible soil and water contamination	
08	Servicing of company cars	Exhaust emissions	Reduction of atmospheric contamination	
09	Replenishment of oil-fired space heaters	Possibility of accidental spillage of fuel oil	Possible soil and water contamination	
10	Packaging	Reuse of packaging material	Conservation of natural resources	
11	Engraving	Plastic and metallic airborne fibres and dust	Possible atmospheric contamination	
12	Engraving	Use of air compressor	Noise pollution	
13	Soldering PCBs	Use of lead-based solder paste	Possible atmospheric contamination	
14	Assembling PCBs	Use of air compressor	Noise pollution	
15	Heating oil storage	Use of fuel tank bund	Possible soil and water contamination	
16	Inherited waste oil tank	Waste oil tank leakage	Soil and possible water contamination	
17	Inherited oil-contaminated soil	Soil contamination	Soil and possible water contamination	

Figure 9.16 Register of aspects and impacts.

Table 9.2 Register of environmental legislation and regulation

Item	Regulation	Regulator	Dated	Applicability
01	Environmental Protection Act	EA	1990	Corporate policy
02	Environment Act	EA	1995	Corporate policy
03	Control of Pollution Act	EA	1974	Corporate policy
04	Waste Management Licensing regulations	EA	1994	Waste recycling
05	Producer Responsibility Obligations (Packaging Waste) Regulations	EA	1997	Reuse of packaging
06	Special Waste Regulations	EA	1996	Chemical disposal
07	The Environment Protection (Prescribed Processes and Substances) Regulations	EA	1991	Corporate policy
08	Noise at Work Regulations	HSE	1989	Air compressor
09	Health and Safety at Work, etc. Act	HSE	1974	Corporate policy
10	Health and Safety (Consultation with Employees) Regulations	HSE	1996	Corporate policy
11	Safety Representatives and Safety Committees Regulations	HSE	1977	First aid representatives
12	Workplace (Health, Safety and Welfare) Regulations	HSE	1992	Corporate policy
13	Electricity at Work Regulations	HSE	1992	Use of equipment
14	Provision and Use of Work Equipment Regulations	HSE	1992	Use of equipment
15	Pressure Systems and Transportable Gas Containers Regulations	HSE	1989	Operation of air compressor
16	Management of Health and Safety at Work Regulations	HSE	1992	Corporate policy
17	Health and Safety (First Aid) Regulations	HSE	1981	First aid requirements
18	Reporting of Injuries, Diseases, and Dangerous Occurrences Regulations	HSE	1995	First aid reporting

Note
EA, Environment Agency; HSE, Health and Safety Executive.

If it helps clarification, identify the positive and negative impacts with a plus or a minus sign.

Register of environmental legislation

It is not unreasonable to say that, of all the forms and documents within the ISO 14001 standard, the NCR and the CAR documents are probably top of assessors' lists for those most frequently and diligently inspected. The aspects and impacts register and the environmental legislation register (Table 9.2) would certainly be a very close second. If you think about it logically, these two registers represent the core of the standard. The register of environmental aspects and impacts is the guide to how your company's operational activities affect the environment. The register of environmental legislation acts as a guide to the types of legislation that could have a negative impact on a company's operations if not carefully monitored.

As environmental manager, you may need assistance when building the legislative register, particularly if you are unsure of the legislation and how it may affect the company. Chapter 4 and Appendix 2 provide information as to the types of legislation and the sources of assistance to identify legislation that are applicable to your company. You will find when compiling the register that the legislation will not be all environmental; there will be some health and safety legislation to be included as well.

Compilation of the register is straightforward. Identify the piece of legislation that affects the company in the first column. In the second, identify whether it relates basically to the environment or health and safety, and note the date that it came into force. In the final column, state to which part of the system or company the legislation applies. Applying this method assists you, or the relevant person, to become familiar quite quickly with the legislation and the areas of the business that may be affected. Like the register of aspects and impacts, environmental legislation should be monitored continually and any new legislation affecting the company should be added as soon as it is approved.

You will notice in the register that some pieces of regulation apply to corporate policy and others to particular activities or departments. Corporate policy indicates that the regulation applies to all aspects of the company's operations such as health and safety and control of pollution. The noise at work regulation may only apply to one aspect of the company's operations, such as the use of air compressors.

10 Manuals, folders and logs assistant

Within the paperwork requirements for ISO 14001, and in addition to the number of documents that need to be produced, there will be a requirement to produce one manual, the environmental management manual, and a number of folders and logs. This chapter provides details on the structure and uses of the manual, folders and logs required.

Manuals

The most obvious starting point is the environmental management manual. If you are a stickler for everything being in its right place, the creation of this manual is for you. If you are not, get someone who is. The paperwork needs to be organised; if it is, you will be amazed at how simple the whole process can be. If a technical author can be recruited into the team, do so; not only will it save a lot of time, but having someone who is experienced with document and form creations and manual and procedure cross-referencing is a definite advantage.

Essentially, there are three options when it comes to creating the framework for an environmental management manual:

1 Follow the ISO 14001 guidelines to the letter. If the manual is structured to the clause statements in the ISO guidelines, it is possible to demonstrate clearly to the assessor that each of the requirements in the standard has been addressed.
2 Follow the plan structure in this book.
3 Structure the manual on the basis of a quality system that is currently operating in the company, or one that you are familiar and comfortable with.

Whichever option you choose, remember that it must suit what the company does and, of course, represent its size. The words hammer, sledge and nut should not be heard in the same sentence when someone is referring to the structure of your environmental manual.

The environmental management manual

The environmental management manual will not hold everything relating to the environmental management system. It will act as your bible, or at least a major reference book for how the system should operate. For those pursuing ISO 14001, it is another piece of evidence to demonstrate that the EMS should be run in accordance with the guidelines laid down for the achievement of ISO 14001 accreditation. In other words, it is the skeletal frame of the system.

The environmental management manual is the single most important document that will be created. It is the focal point of the system from which all other relevant documentation and procedures are referenced. When constructing the manual, keep it clear and concise; this is where the technical author's skills come to the fore. As well as a reference document, it can also be used as an excellent training tool to make existing employees aware of the way(s) in which the system conforms to the ISO 14001 standard, and to supplement the induction training for new employees with some additional procedural information.

ISO 14001 specifies that a document must be produced that describes the main elements of the EMS. Where it is integrated with other management systems, such as health and safety and quality, the scope of the environmental management manual should also include references to these systems and procedures and show how they interrelate. Do not include material in the environmental management manual that is likely to change often because this will result in frequent and undesirable reprints and reissues.

Folders

The next stage down from the environmental management manual, in the paperwork hierarchy, is the folder. The function of the folder is to control the operation of an element of the environmental management system. The environmental management system and how it operates is detailed in the environmental management manual. The evidence of how the environmental management systems are controlled is kept in the following folders:

- environmental correspondence folder;
- environmental documentation folder.

Environmental correspondence folder

The ISO 14001 system requires evidence to prove that the system is doing what you say it is doing. To achieve this effectively, you must communicate, both internally and externally, with other parties that may have an input into the system. Any internal communication concerning changes to the system or notification of meetings would be kept in this folder.

There are many external parties, such as the water regulator, local council, environmental agency, waste contractor and suppliers, with whom there will be regular environmental correspondence. These communications are important pieces of evidence that need to be kept in the folder to be called upon for verification, justification or ISO 14001 assessment purposes. The environmental correspondence folder would have, as a minimum requirement, eight sections for key correspondence:

- local council;
- suppliers;
- environment agency;
- agreements;
- assessors' reports;
- internal memorandums;
- company drainage plan.

Local council

The extent of correspondence with the local council will depend on the company's level of involvement with them. The council may become involved in the removal of any major scrap items, such as freezers, fridges and cookers. It would also be involved with company rubbish collection and the provision of 'wheelie bins'. If the company owns the company premises, the local council's involvement should be minimal.

If the company leases premises from the local council, there will be a greater involvement by the council. The main reason for this is that any proposed environmental changes, as identified in the aspects or impacts analysis, may have to be ratified by the local authority before they can be actioned. Although this situation has the potential to raise barriers to the implementation of the ISO 14001 system, if handled correctly these barriers can be easily avoided. The case study in Box 10.1 demonstrates a major barrier that may occur between a company and the local council and how it may be overcome.

Most local councils are very supportive of any improvements that a company may wish to make to its premises. A good starting point for discussion with any local council is to write to the estates office and advise it of the environmental management system being implemented. You should list the remedial actions that the company intends to undertake and list those points that the council is required to deal with. A constructive working relationship should follow quickly.

Box 10.1 Inherited problems

> A small manufacturing company leased its premises from the local council. The lease was for a fixed period of time on a full insuring and repair and maintenance basis. In other words, a standard business lease agreement.
>
> When setting up ISO 14001, the company found that it had a large tank full of waste oil at the rear of the building and the heating oil storage bund contained 30 cm of fuel oil-contaminated rainwater. To add to its problems, the contents of the bund were leaking through the wall, and the ground outside the bund was also contaminated. The cost of cleaning up the contaminated water and soil and the disposal of the waste oil was calculated. The proposed cost of the clean-up was high and was prohibitive in relation to the size of the company.
>
> After discussions with various environmental experts, it became clear that the cost of paying for the clean-up was not the company's responsibility because the waste oil and contaminated soil and water were inherited from a previous tenant.
>
> Subsequent correspondence with the local council revealed that the owner of the land is responsible for the clean-up of any contamination of land or water. This outcome, following an exchange of letters with the local council, ensured that the manufacturing company attained ISO 14001 accreditation without having to spend huge sums in clean-up costs.
>
> These days, most if not all company solicitors would undertake surveys to guard against such occurrences. These surveys, however, may not have been carried out as much as 10 years ago and those companies will long-term leases may well find themselves in a situation like this.

   Hint

Most responsible ISO 14001 accreditation companies will ensure that a company does not spend huge sums on unnecessary environmental improvements. When installing, or improving, the environmental management system, the acronym to bear in mind is BATNEEC: *b*est *a*vailable *t*echnique *n*ot *e*ntailing *e*xcessive *c*ost.

Suppliers

Evidence of correspondence with suppliers is also a requirement. Part of achieving the ISO 14001 standard is the monitoring of business activities, i.e. what impacts the company's products and processes have on the environment. Another part is the monitoring of the components used within these products and processes, which means checking that suppliers are providing the necessary information about the items or components they are supplying to your company. Again, parsimony is the key word here. It may be necessary to monitor just a few suppliers: those that are major suppliers to your company and those that supply particularly hazardous items such as chemicals or solvents (Figure 10.1).

The most efficient way of obtaining the information that you require is to devise a questionnaire with a few key questions that can be completed quickly by the supplier. If any suppliers have the ISO 14001 standards themselves, they would do this automatically. A questionnaire format is provided in Chapter 9 as an example; you can, of course, create your own to suit your own company's requirements.

Environment agency

Wherever the company is located, there will be an environment agency responsible for environmental management for your area. If you are unsure of the agency for your area, see Appendix 3 for details. Your local environment regulation agency is a good first port of call for any environmental query. If it cannot help you, it will probably know where to direct you.

An important task of any environmental agency is to monitor and issue licences to waste carriers to dispose of industrial waste. The agency maintains a register of all approved waste carriers, their names, addresses, contact numbers and licence numbers, and the waste types for which they are licensed (Figure 10.2). You will need this information to organise how, and when, the company's waste is to be transported and disposed of. Bear in mind that the licences issued for waste disposal and waste carrying are different; you may need to check for more than one licence.

Figure 10.1 'No Chalmers, I said we need to grow a culture of parsimony'.

Company name	Address	Contact details	Licence no.	Waste category

Figure 10.2 Register of waste companies.

 **Hint**

You will receive a copy of the waste disposal company's licence, which will come straight from them; it is worth contacting the Environment Agency to ask for confirmation that the licence is still current.

Agreements

The installation of an environmental management system will invariably require agreements to be made with various agents. Evidence of these agreements is, as with all elements of the ISO 14001, important for the assessor's peace of mind.

As an example, your company will have agreements with the local council or a waste disposal company for a trade waste collection service. This agreement may come in two parts: first, a notice of charges for the provision of bins and for skips will be provided, giving a breakdown of the various elements of the cost of hire by size; second, a 'waste transfer note' will be supplied. The waste transfer note is one of the first pieces of paper that any reputable assessor will ask for when conducting an interim assessment, a final assessment or follow-up audits after the environmental management system has been successfully installed.

 **Hint**

For ISO 14001, assessors always, without exception, check that all the company's licences and agreements are valid. Most only last for 1 year, others less. It is easy to forget to renew licences and agreements; the onus is on you to update them, not on the agent.

If the company disposes of other waste such as scrap metal, wood, plastic or cardboard, the local council may not be able assist, and therefore the company may have to depend on a private waste management company to take the waste away. Another agreement and, yes, another charge, but it should be no more than £3–5 per week, depending on the amount of waste to be removed. Remember that the duty of care for the disposal of waste is on you. You must ensure that the private waste carrier has the necessary licences and that they are current. It is also very important to know where the company's waste will eventually end up and how it is handled.

  Hint

If an unlicensed waste carrier takes the company's waste away, the company will remain responsible for that waste irrespective of where it is finally disposed of or how it got there. The waste remains the company's until another party accepts it and issues to the company a valid duty of care certificate to say that the waste company will dispose of the waste in a responsible manner. Also, if an unlicensed waste carrier removes the waste, you – as environmental manager – will be committing a crime for which you can be prosecuted.

Accreditation reports

By far the largest section in the environmental correspondence folder will be the reports and other correspondence with your chosen accreditation company. Whichever company you choose, the paperwork will be significant and will begin with the registration document followed by interim assessment reports and fee notices. It will culminate with the final assessment, which recommends the award of the environmental standard that you are seeking. Other paperwork in this section may include clarifications of environmental law or process changes that affect your business's activities.

Internal memorandums

It is worth mentioning that creating a section for all internal environmental correspondence will pay future dividends. There will be times when reminder, or action, memorandums relating to the environmental management system will be required. Some of the corrective action requests may be achieved by memorandum; these all need to be kept as evidence of actions taken. This section may also contain the questionnaire that was originally presented to the workforce from which some of the environmental aspects of the company were determined. The assessor will ask to see this questionnaire and the responses to it at most, if not all, of the assessment stages.

Local water authority

In some areas, water pollution is handled by the local water authority and not by the Environment Agency. If this applies to your company, any water discharges from its premises will need the approval of the local water authority. If your business's activities or processes use water and then discharge this to the sewers, storm drains, land or rivers, a sample of this discharge will need to be analysed by the local water authority to ensure that it is within acceptable limits.

If, for example, the company's operational activities involved plating metals with either tin or copper and the residue of this operation was discharged to the sewer, the discharge would obviously contain traces of heavy metals. The results of a tested sample would be along the lines of the sample test report shown in Table 10.1. This will identify all the heavy metals contained in the sample, and the number of micrograms of each metal found in the sample. The results are given as micrograms per litre, but you are not advised of the minimum acceptable levels. You may need to ascertain the minimum acceptable levels separately because they are not normally provided as part of the test report.

Depending on the volume of discharge, a 'trade effluent consent' may also be required. If an assessment is made of the volume of the water discharge, the sewage undertaker/water authority will decide whether a trade effluent consent is required. The likely outcome is that a visit will be arranged and a judgement will be based on the results of the sample and an estimation of discharge volume. Incidentally, there is a charge for having the sample tested; this is somewhere in the region of £100 per sample, but it may vary from region to region. An accredited laboratory can carry out the analysis, so you do have a choice.

A visit from the local sewage undertaker/water authority will confirm the report details that you have provided concerning the volume and content of what your company is discharging. The authority will also, using coloured dyes, check the drainage plan that you have provided to ensure that the direction of effluent flow is correct. If all discharges are going to the sewer, this can generally be regarded as good and, barring other considerations, a discharge consent will be issued.

Table 10.1 Test report analysis

Element	Result	Units	mg/litre limit
Copper	20	µg Cu/l	5,000
Zinc	138	µg Zn/l	5,000
Cadmium	10	µg Cd/l	500
Lead	360	µg Pb/l	5,000
Chromium	50	µg Cr/l	2,000
Iron	200	µg Fe/l	5,000
Nickel	30	mg Ni/l	5,000

Company drainage plan

A drainage plan is required at the very beginning of introducing an environmental management system. If there are emergency or accidental discharges of substances resulting from your business's activities, you, and other interested parties, will need to know where the substance flows and where it is likely to end up. The local council planning office may be able to supply you with a copy of the drainage plans for your company's premises. They may charge a nominal photocopying fee for the privilege. If they do not have plans of your company's premises, and it has been known to happen, then draw up your own. The plans do not need to be to scale, but they do need to show the layout of the premises and the locations of those processes that use water (including toilets) as well as internal and external pipework, drains, downpipes, sewage and storm drain manholes. An indication of the direction of flow is also required. If there are different water uses, it would be useful to colour code the flow to the different discharges (Figure 10.3).

Environmental documentation folder

This is the folder where all of the environmental documents or forms necessary for the correct functioning of the environmental management system will be kept. Although this folder holds all of the blank original forms, it is also useful to store completed forms, such as audit reports and duty of care certificates. It is easier and more efficient if all forms, blank or

Figure 10.3 'Drink this blue water Watkins so we can see the direction of flow for the drainage plan.'

completed, internal or external, are kept in the same location. Samples of these forms are shown in Chapter 9.

Logs

The next stage down again in the paperwork hierarchy is the humble log. The log provides a chronological representation of various actions that have happened within the system. Working on the basis of parsimony, but also satisfying the ISO 14001 standard requirements, six key logs are recommended for use:

1 environmental non-conformity report (NCR) status log;
2 environmental corrective action request (CAR) log;
3 environmental document log;
4 environmental communications log;
5 waste collection log;
6 audit log.

Environmental non-conformity report (NCR) status log

An NCR is a non-conformity report. You will use, and become very familiar with, these phrases as you progress through ISO 14001 accreditation. A non-conformity is an activity, or function, of some part of the environmental management system that does not do what the procedures say it should do.

When a non-conformity is recognised or discovered, it is presented in a report so that it can be rectified. Do not be alarmed though – non-conformities can be good things. When they occur and are corrected, then you can be sure that the system works; also, if you are going through the accreditation process, you can be sure that you are making progress. All of the NCRs raised need to be recorded in the environmental non-conformity report log, which provides an up-to-date status report of all the corrections made to the system (Figure 10.4).

NCR serial no.	Date NCR raised	Review meeting date	Brief description of non-conformity (including category)	Brief description of agreed corrective/ preventive action(s)	CAR serial no.	Date CAR raised

Figure 10.4 NCR status log.

Major non-conformities

A brief word on major non-conformities will be useful at this point. The only non-conformities to avoid are major non-conformities; these, if encountered, will cause a company to fail an assessment immediately. Installing ISO 14001 and going through the assessments and final accreditation is similar to the board game 'Mousetrap'. You piece it all together painstakingly, making sure you have a sound foundation and you keep introducing all the changes. You make mistakes along the way, but this is inevitable and should not deter you. You will forget some piece of paperwork or not recognise a particular aspect or impact of the business's activities. If the assessor finds these errors, the company will receive a non-conformity assessment, which will be flagged for correction. This resembles the game of 'Mousetrap'; if a piece falls over, it may delay progress but the trap does not fall.

A minor non-conformity, for example, would be to omit the identifying numbers from the documents used in the EMS. A major non-conformity, by contrast, would be to discharge polluted water into a nearby stream, river or other waterway. Apart from bringing the implementation process to a juddering halt, such a major non-conformity would carry a hefty fine.

If, for example, effluent was discharged to an open drain and not recognised, the outcome would be a major non-conformity. If the failure was recognised by your company's assessor during any stage of the assessment, the process would stop immediately. The mousetrap would most certainly fall and the game would be lost. You would only be able to play again when the major non-conformity had been rectified, and the rectification had been notified officially to the assessor.

In short, then, if non-conformities occur and are dealt with quickly and correctly, they will gradually improve your company's environmental management system and it will eventually receive accreditation. Major non-conformities, unlike minor non-conformities, stop assessments or accreditation in their tracks. To continue the board game analogy, you will have to start the game again. You will waste time and money as the failed assessment will have to be repeated. Avoid this at all costs as it can be very demoralising for the whole company.

Environmental corrective action request (CAR) log

When an NCR has been raised and logged, the cause of the NCR must then be corrected. The instrument used for correction is the corrective action request (CAR).

The CAR is the solution to the NCR. Each NCR raised can only be closed out by a CAR. This means that, once the non-conformity has been identified, its remedy is detailed in the CAR and the person responsible for its action ensures that the details of the remedy are carried out. Once the CAR has been actioned, it is recorded in the CAR log (Figure 10.5) as being complete.

Audit no.	CAR no.	Date raised	Originator	Addressee	Brief description of non-conformity (including category)	Corrective action					CAR close-out
						Recommended		Completion			
						Due	Received	Due	Verified		

Figure 10.5 CAR status log.

The corrective action log provides a quick reference to all corrective actions raised, those that have been completed, those that are in progress and those that are still to be actioned.

Environmental document log

Every manual, folder, log, document form, matrix, note, report, etc. required for ISO 14001 will have a document reference number. These can all be numbered sequentially. It is up to you whether the documents are numbered alphabetically or as they are completed. One recommendation would be to have three letters and three numbers, for example ENV001. Why? Because you may have other systems with similar referencing styles. A good, clear numbering system will avert any possible paperwork confusion. For larger companies, the addition of an 'AA' reference, particularly if you are identifying a number of sites, departments or companies within a group, would give greater flexibility. For example, ENV/001 would become ENV/AA/001.

Whatever system you choose, ensure that each document has its own unique number. As well as the document number, the document log (Figure 10.6) will also require an issue date and a revision number, which should be carried on every page of each document. The document issue date shows when the document was created and the revision number shows how many times the document has been updated or changed. As the system matures, you may find that a number of revisions are required to some documents, particularly as the system moulds itself to the way that the company operates.

Environmental communications log

The environmental communications log (Figure 10.7) is quite simply a record of all external and internal correspondence regarding the system. We discussed earlier in this chapter the function of the environmental correspondence folder. The communications log provides a chronological account of the correspondence undertaken and the medium by which it was created, e.g. letter, fax, e-mail. If your company has a register for recording incoming and outgoing mail, this works on exactly the same principle but relates specifically to the environmental management system.

Waste collection log

Within one of the operating procedures for waste handling and control, there will be details on how waste is collected, stored and disposed of. The definition of waste may include paper, plastic, wood, metal, wire, printer cartridges, etc. The type and volume of waste items will vary from business to business and across industry sectors. Whatever your company

Environmental document or form title	Document no.	Issue	Revision
Environmental management manual	ENV000		
Waste collection log	ENV001		
Environmental non-conformity report (NCR)	ENV002		
Environmental non-conformity report (NCR) status log	ENV003		
Corrective/preventive action request (CAR)	ENV004		
Corrective/preventive action request (CAR) log	ENV005		
Document transmittal advice	ENV006		
Document log	ENV007		
Training schedule	ENV008		
Document distribution matrix (EMM)	ENV009A		
Document distribution matrix (EDF)	ENV009B		
Document distribution matrix (ECF)	ENV009C		
Supplier environmental questionnaire	ENV010		
New process introduction form (NPI)	ENV011		
Communications log	ENV012		
Correspondence folder	ENV013		
Documentation folder	ENV014		
Organisational chart	ENV015		
Training syllabus – induction	ENV016A		
Training syllabus – refresher	ENV016B		
Duty of care – controlled waste transfer note	ENV017		
Audit notification	ENV018		
Management review agenda	ENV019		
Audit log	ENV020		
Audit report	ENV021		

Figure 10.6 Documentation log.

Date	Time	Message type	Message from (name, company, etc.)	Message details	Follow-up action

Figure 10.7 Communications log.

classifies as waste has to be disposed of, whether by the company or by an external contractor, in a responsible manner, remember the 'duty of care'. To ensure that there is the necessary evidence of the way in which the waste was disposed of, a waste log (Figure 10.8) is required.

There are some additional benefits to keeping a waste log other than the recording of the different quantities of waste disposed of on a monthly basis. Keeping a regular check on the quantities of waste being disposed of can offer additional insight into inefficient operational activities or processes. If the quantities are large enough, additional business opportunities may arise for reusing or reprocessing the waste to generate additional income or reduce costs.

The waste log in itself is sufficient to record the quantities of waste leaving the business. The preparation of a graph of waste quantities, however, adds to the quality of information and can help in waste trends identification, prompting any remedial action that may be required (for more details on the waste log, see Chapter 9).

Audit log

The audit log (Figure 10.9) provides a comprehensive record of personnel who have been audited in accordance with the requirements of the environmental management system. The process of recording audits and thus identifying which members of staff have or have not been audited previously has two benefits – the more people you can audit the more confident you will be about passing the final assessment and the more people you audit the more familiar they become with the requirements of the EMS. Take the opportunity wherever and however it presents itself to train employees and to reinforce the environmental management system.

Environmental legislative update folder

The creation of a folder to store the source of information used to keep

| Date | Waste type and estimated quantity (kg) | | | | | | | Collected by | Signed |
	Paper	Cardboard	Plastic	Non-ferrous	Ferrous	Rubbish	Other		

Figure 10.8 Waste collection log.

Auditee	Audit date	Auditor	Auditor signed
Employee 1	28 March 2001	I. Crane	
Employee 2	28 March 2001	M. Minor	
Employee 3			
Employee 4			

Figure 10.9 Audit log.

you and your company's system up to date on changes in environmental legislation is an ISO 14001 necessity. The register of environmental legislation has to be kept up to date. Any changes in existing legislation or the introduction of new legislation that may have implications for your business's activities need to be monitored, and the records updated as appropriate.

This may seem at first glance to be another needless paper exercise. If used correctly, however, it can provide forecasting opportunities that will allow the company sufficient time to discuss various potential courses of action to be taken before the legislation/regulations come into force. All too often, companies find out about legislative effects as they happen, and time and money can be lost in making rapid, and often ill-considered, operational adjustments.

There are many environmental information companies that offer information services, some of which are listed in Appendix 2, and there are many media such as magazines and Internet bulletins from which the information can be gathered. Whichever medium you choose, check that the provider has the necessary environmental expertise and offers at least monthly updates. As always, there is a cost for such services, but it should not be more than £150–200 per annum. During assessment, it will be necessary to produce evidence of the information source used, so if it does not have its own folder, and most do, create one.

Document controlling and updating

Certainly, you will find that maintaining a concise array of manuals, folders, logs, etc. will help the system to operate more smoothly. To ensure that it remains smooth, a continuous updating process must take place. This does not need to be carried out on a daily or even a weekly basis. When conducting the monthly audits, it is worthwhile considering quickly those manuals, folders and logs that may need updating, particularly when changes have been brought about through non-conformities, or product or process changes. Whenever changes, or revisions, to the procedures or system are made, they are required to be recorded in the revision record at the front of the environmental manual. Details of how these changes are recorded are provided in Figure 10.10.

Depending on the number, or impact significance, of procedural or other changes, it will be necessary to decide at some point when to reissue the entire environmental manual with all revisions included. You will be the best judge of when to do this. If it is left too long, however, the assessor may influence this by raising an NCR.

When trawling through the paperwork requirements of ISO 14001, you will stumble over the term 'controlled document'. Once you and the assessor have agreed that the environmental management manual conforms to the requirements of the ISO 14001 guidelines, it is essentially cast in stone. The environmental management manual is then issued as 'issue 1'. You will determine who will keep the master copy and who will receive additional copies. As demonstrated in Figure 10.11, a record is kept of the number of copies issued and the recipient of each copy. These are the controlled documents.

Once the decision has been made as to who will receive a copy of the environmental management manual, it is probably best to keep the numbers down to save on printing costs. A record must be kept of all copies issued. A document is created similar to the one shown in Figure 10.10 that identifies which document is being distributed. It should also list the recipients, names, their locations within the company, and the issue date and revision number (if applicable).

If there are many changes to the contents of the manual, eventually the whole document will need to be reissued. When this happens, the whole process is repeated, this time with a different issue number. Folders, logs and registers are not classified as controlled documents in the ISO 14001 standard and, as such, do not have to be monitored in their distribution. Of all the manuals, folders and logs identified earlier in this chapter, the

Document title: Environmental management manual			
Status	Controlled document number and authorised person 01 – Master copy D. Brown	02 – Office P. Green	03 – Workshop O. White	04 – Engineering M. Pallett
Issue	April 2001	April 2001	April 2001	April 2001
Revision	00	00	00	00

Figure 10.10 Environmental management manual (EMM) revision record.

Document title: Environmental management manual			
Status	Controlled document number and document holder 01 – Master copy 02 – Office 03 – Workshop A. N. Other O. B. Kenobi L. J. Silver		
Issue	June 2001	June 2001	June 2001	
Revision	00	00	00	
Issue				
Revision				

Figure 10.11 Controlled document reference.

environmental management manual and the environmental correspondence folder are controlled documents.

Issuing documents

Every controlled document has a control number. Each copy of the
document sent out has a control number. These numbers are required to
be displayed clearly on the cover or spine of each document. As you hand
over the treasured document to a colleague, you will need a receipt, again
as evidence. Together with each controlled document, a document
transmittal advice must be presented for signature.

Document transmittal advice

The document transmittal advice is an official piece of paper that proves
that you – as environmental manager – are aware of the location of the
environmental management system documentation. Each recipient of an
environmental management manual will sign for a copy, or any updated
contents deemed necessary for the document. The signed receipts are
returned together with the superseded paperwork, or environmental
management manual, to the person responsible for the running of the
system.

11 Alternative environmental management systems

If you have committed yourself to this point in learning about environmental management, you will have mastered the basics for putting together an environmental management system that suits the way your company operates. The focus of this book is on the successful implementation of ISO 14001. This chapter seeks to extend your knowledge of environmental management systems and to offer an alternative EMS if you consider that ISO 14001 is not appropriate for your company.

Some background

The history of the development of environmental management systems will provide an insight into the road that has been travelled, and the experiences learned along the way, by others. The majority of large companies have environmental management systems and environmental policy statements declaring that their objective is to improve environmental management and reduce exposure to environmental risk.

Most managers' earlier views of environmental management systems were based on the assumption that the systems were detrimental to the principal managerial goals of profitability, maintaining market share, cost control and production efficiency. These views were based on three fundamental assumptions. The first was that management believed the benefits of following sound environmental practices could not be achieved because consumers were not prepared to pay for the increased costs to industry. Second, the business costs of environmentally sound strategies were significant. Third, industry was not prepared to have an increasing investment in non-productive areas such as environmental risk aversion.

Today, successful companies demonstrate that previous negative environmental management systems have proved ineffective in a dynamic, rapidly changing business environment. Many companies have since adopted a more positive and creative approach to environmental management

By addressing growing public concern for environmental sensitivity and the legislative requirements of industry, companies are gradually becoming more aware of the benefits of having an environmental focus. Research shows that there is profit in going green and that sound environmental management can lead to competitive advantage in business practice.

To address the issue of environmental performance in industry, environmental standards were considered to be the best way forward in demonstrating the benefits of better environmental management to industry and to offer a process, similar to the familiar BS 5750 quality standard, to reduce the risk of exposure to the increasing numbers of European environmental directives.

BS 7750

BS 7750 was the first UK national standard created for an environmental management system. Based on the BS 5750 quality system, the BS 7750 system was used to describe the company's environmental management system, evaluate its performance and to define policy, practices, objectives and targets; this provides a catalyst for continuous improvement.

The concept is similar to that of BS 5750 and ISO 9000 for quality systems, in which the methods to be used are open to definition by the company. The standard provides the framework for development and assessment of the BS 7750 environmental management system. BS 7750 was developed as a response to concern about environmental risks and damage (both real and potential). Compliance with the standard is voluntary and complements the requirements for compliance with statutory legislation. As BS 5750 was the driver for ISO 9001, so BS 7750 led to the development of ISO 14001.

As its base, BS 7750 requires an environmental policy to exist within the company that is fully supported by senior management and that outlines the policies of the company not only to the staff but also to the public. The policy needs to clarify compliance with environmental legislation that may affect the company and stress a commitment to continuous improvement. Emphasis has been placed on policy as this provides the direction for the remainder of the management system.

The preparatory review and definition of the organisation's environmental effects is not part of a BS 7750 assessment, but examination of these data will provide an external auditor with a wealth of information on the methods adopted by the company. The preparatory review itself should be comprehensive in its consideration of input processes and output at the site. It should also be designed to identify all relevant environmental aspects that may arise from the company works. These may relate to current or future operations, as well as to the activities performed on site in the past, e.g. contamination of land.

EMAS

The Eco-Management and Audit Scheme (EMAS) is similar in structure to ISO 14001. There are, however, two major differences between the standards. The first is that the whole company can be certified to ISO 14001, whereas EMAS is generally a site-based registration system. The second is that, whereas any company from any business sector can use ISO 14001, EMAS is only available to those companies operating in the industrial sector. If you are in doubt as to your company's industrial status, contact an accreditation company and they will be able to advise you (see Appendix 3). Within the UK, an extension to the EMAS scheme has been agreed for local government operations, which may also register their environmental management systems to the EMAS regulations.

In addition to a summary of the process, the statement requires quantifiable data on current emissions from the site and environmental effects, the amount and types of waste generated, raw materials utilised, energy and water resources consumed, and any other environmental aspect that may relate to operations on the site.

Preassessment is as much part of EMAS as it is of ISO 14001. The environmental audit must be comprehensive in consideration of input processes and output at the site. The procedure is designed to enable identification of all relevant environmental aspects that may arise from the site itself.

The preassessment will also include a wide-ranging consideration of the legislation that may affect the site, whether it is being complied with currently, and perhaps even whether copies of the legislation are available. Many of the environmental assessments undertaken already have highlighted the fact that companies are unaware of all the environmental legislation that affects them, and, being unaware, they are often found not to be meeting the requirements of that legislation.

Under the EMAS standard, the company will declare its primary environmental objectives, i.e. those that can have most environmental impact. In order to gain most benefit, these will become the primary areas of consideration within both the improvement process and the company's environmental programme. The programme will be the plan used to achieve specific goals or targets along the route to a specific goal and will describe the real and achievable means to be used to reach those objectives.

As with ISO 14001, the EMAS standard requires a planned, comprehensive and periodic series of audits of the environmental management system to ensure that it is effective in operation, is meeting specified goals, and continues to perform in accordance with relevant regulations and standards. The audits are designed to provide operational information in order to exercise effective management of the system, providing information on practices that differ from the current procedures or offer an opportunity for improvement. Under EMAS, the bare minimum frequency for an audit is every 3 years.

Most companies are used to producing an annual report and accounts describing the activities of the organisation over the previous year and its plans for the future. EMAS generally requires a similar system for the company's environmental performance. It is also a requirement that there should be a statement about performance during the previous period, a set of current performance data to be achieved and notice of any particular plans for the future that may have an effect upon the environmental performance of the organisation, whether detrimental or beneficial.

The peculiarity with EMAS is that the policy statement, programme, management system and audit cycles are reviewed and validated by an external, accredited, company. In addition to providing a registration service, this company is also required to confirm, and perhaps even sign, the company's periodic environmental statements.

Environmental management system standardisation

In 1990, the British Standards Institution (BSI) in the UK devised an environmental assessment standard based on the then existing quality standard BS 5750, now superseded by the ISO 9000 series of standards. After consulting industry, and following a 2-year pilot programme, in 1994 BSI launched the environmental standard BS 7750.

At the same time, the European Commission put forward a proposal for a scheme known as the Eco-Management and Audit Scheme (EMAS). The objective of this was to put the emphasis on the management of environmental systems not just on the system itself. EMAS was launched in 1995.

The BS 7750 and EMAS standards were very similar in environmental requirements for businesses. At the time, it was thought appropriate to make EMAS a mandatory standard for businesses. Strong industry lobby groups argued successfully, however, that a mandatory approach would be detrimental to industry and EMAS is now a voluntary scheme.

In 1993, it was felt that an international standard was required for environmental management. Three years later, in 1996, from an idea based on BS 7750, the ISO 14001 standard was born. As BS 5750 had been withdrawn with the appearance of the ISO 9000 series, the outcome of the emergence of the international standard ISO 14001 meant that the national standard BS 7750 – together with national standards in other EU countries – were, with common consent, also withdrawn (Figure 11.1).

Since the introduction of ISO 14001, many other ISO 14000 standards have also come into operation, as follows.

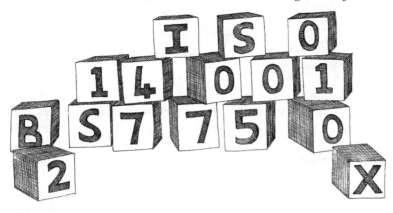

Figure 11.1 ISO 14001 is based on BS 7750.

ISO 14001 standards

Standard	Description
14000	Guide to environmental management principles, systems and supporting techniques.
14001	Environmental management systems – specification with guidance for use.
14010	Guidelines for environmental auditing – general principles of environmental auditing.
14011	Guidelines for environmental auditing – audit procedures, part 1: auditing of environmental management systems.
14012	Guidelines for environmental auditing – qualification criteria for environmental auditors.
14013/15	Guidelines for environmental auditing – audit programmes, reviews and assessments.
14020/23	Environmental labelling.
14024	Environmental labelling – practitioner programmes – guiding principles, practices and certification procedures of multiple criteria programmes.
14031/32	Guidelines on environmental performance evaluation.
14040/43	Life cycle assessment general principles and practices.
14050	Glossary.
14060	Guide for the inclusion of environmental aspects in product standards.

Acorn project

The most recent development designed to assist SMEs to achieve ISO 14001 is a trial project called Acorn. This is a pilot project focused on introducing greater flexibility to those SMEs seeking to achieve the ISO 14001 standard.

Launched in the summer of 2000, Acorn is a joint Department of Trade and Industry (DTI) and BSI pilot study to test a five-step model for SMEs to achieve ISO 14001.

The five steps are as follows:

Step 1   A commitment to the ISO 14001 standard.
Step 2   Compliance with legal and regulatory requirements.
Step 3   Identification of significant environmental aspects and impacts.
Step 4   Management of significant environmental aspects.
Step 5   Documentation and integration of environmental management system.

As major companies are coming under greater pressures to ensure that their suppliers also have the ISO 14001 standard, the Acorn project is designed to assist SMEs in meeting major company environmental demands. SMEs do not necessarily have to achieve full certification before becoming part of a major supply chain. Stage 3 of the Acorn project is the first validation and audit point for the standard; for most major companies, those SMEs achieving this stage will be accepted as a supplier.

The Acorn initiative allows SMEs with limited resources to achieve the standard in their own time while still adhering to major company environmental requirements – the pilot project will be completed in 2002.

Sustainable development

The ultimate goal for corporate environmental management systems should be to achieve a situation where a company's activities are in balance between continuous economic growth and caring for the environment. The International Chamber of Commerce (ICC) has developed a business charter for sustainable development. This effort has the support of many companies and is intended to identify areas for successful and continuous improvement. The charter promotes environmental policy areas as follows.

Corporate priority

The aim here is for environmental management to become one of the highest priorities for all companies. This should be demonstrated by the creation of policies, programmes, procedures and practices leading to environmentally sound operations for the company.

Integrated management

Full integration of environmental elements into the normal business operation of a company is essential to ensure that the management system functions as a coherent entity. The strength of this integrated approach can be demonstrated further by amalgamation of the management systems for environment, quality, customer satisfaction, and health and safety of staff. In effect, environmental management cannot be achieved by one person in a small office at the rear of the company.

The process of improvement

The process of continuous improvement, from a sustainable development viewpoint, requires more than the operations of a company to be environmentally friendly. It requires consideration to be given to how the company responds to customer expectations, community expectations, legal regulations, development of tools and new techniques in order to improve continually the environmental performance of the organisation.

Staff education and training

The development of a culture of continuous environmental education and training will provide greater awareness capability and motivation to employees and enable them to conduct their daily operational activities in an environmentally responsible manner. Such a programme can be a strong motivational tool and can bind employees closer to the company's environmental ethic.

Products and services

The achievement of sustainability requires consideration of the life cycle of products and services (the term 'cradle to grave' is used extensively) to ensure that they have no undue environmental impact and are fit for their intended use, are efficient in their consumption of resources, and can be recycled, reused and disposed of safely.

Facilities and operations

Similar to the 'cradle to grave' concept for products and services, facilities and premises can be renewed, regenerated or upgraded. When looking at existing plant and facilities, a stronger emphasis can be placed on the promotion of renewable resources, minimisation of pollution, the generation of waste and safe and responsible husbandry of the land resource.

Contractors and suppliers

The company should encourage contractors and suppliers to adopt sound environmental principles combined with a system of continuous improvement in line with or supported by its own system. Greening the supply chain ensures that the suppliers' objectives become consistent with the aims of the company itself. This also encourages the wider adoption of environmental improvement practices among suppliers and sub-contractors. The bigger the company, the easier and more readily this concept will be accepted by suppliers and subcontractors.

Openness to concerns

Sustainable development requires a company to become more transparent in all its business operations. This open approach is designed to foster an open exchange of information and communication between the company and its employees, suppliers, stakeholders, etc.

Compliance and reporting

As part of a drive to be more transparent, a company pursuing sustainability will be required to measure and publicise environmental performance information. Following the publication of a company's environmental aspects and impacts, measurements can be made to demonstrate stability or improvement, and to confirm compliance with company objectives and legal requirements. This information is produced annually to provide information to all stakeholders.

12 ISO 14001 self-assessment

If, before deciding on whether to pursue the ISO 14001 environmental standard, you wanted just to compare where your company is now with what has to be achieved, it is worth completing a preassessment questionnaire (see Figures 12.1, 12.2 and 12.3). Before undertaking this exercise, it helps to have a fair understanding of the environmental terminology used and what is likely to be discussed at the preassessment meeting with your appointed assessor. This chapter will help in understanding how a company is assessed, and provide an insight into the assessment process and the route to final certification.

Whichever assessment company you choose to achieve the environmental standard, the preassessment questionnaire should follow a similar format. It will comprise a number of sections relating to your company and how it interacts with the environment. The following list indicates the areas to be assessed:

- the company;
- air emissions;
- water pollution;
- solid and hazardous waste;
- soil and groundwater protection;
- noise control;
- resource management;
- management systems;
- other environmental impacts;
- indirect impacts;
- environmental exercise.

The company

This section requires you to supply general company information, e.g. name, address, number of employees and the nature of the business. State also the company's main business activity. If the company has more than one site, provide the same details for each site. If your company is aiming for the ISO 14001 standard, all company sites must comply with the standard's requirements.

Air emissions

This element asks you to consider which of your company's processes or business activities produce emissions to air. Consider everything vented, from heating fumes to incinerator flue gases, and record it. It is not necessary to be fully aware at this stage of the emissions being vented; this can be set down later by a third party if your company does not possess the expertise in house. You should be aware that, within the preassessment questionnaire, your company will be asked to provide answers to simple questions such as, 'do you monitor air emissions?', 'who is responsible for air emissions?', 'are they audited?' and 'is there a site plan showing emission points?' These are general 'yes' or 'no' questions that will provide the company and its assessor with the basic information needed to consider whether or not air emissions are to be an issue when going for full certification (Figure 12.1).

Water pollution

A business is likely to use a significant amount of water, in everything from toilets to cooling processes. Make a log of everything that uses water. Make special note of those processes or activities that discharge water to the foul sewer, soakaways, other drains or natural streams, etc. With the same tick box 'yes'/'no' response, questions will be asked about wastewater objectives and targets, e.g. 'do you have documented procedures and is there a drainage plan of all water usage and flows?' (Figure 12.2).

Solid and hazardous waste

Your company also has a requirement to identify any waste that it produces. There are usually four main areas:

1   solid waste – e.g. paper, plastic, metal, wood, etc.
2   hazardous waste – e.g. sludge from paper manufacture, paint-spraying residues, etc.
3   special waste – this has a lengthy legal definition that you are directed to if you are unsure of your company's waste classification, and will include chemicals and other substances that are highly flammable, irritant, harmful, toxic, carcinogenic or corrosive.
4   clinical waste – again you will need to reference a legal definition if your company produces clinical waste; put simply, this covers medical waste including instruments and diseased waste.

Type of emission	Activity	Site legislation/regulation/consent

Who is responsible for managing air emissions?	Yes	No
Do you monitor air emissions?	❑	❑
Do you have objectives and targets for air emissions?	❑	❑
Do you have documented procedures for identifying, monitoring and controlling air emissions?	❑	❑
Have you conducted training for air emission procedures?	❑	❑
Have you audited these procedures?	❑	❑
Do you have a site plan illustrating emission points?	❑	❑
If so, where is this plan kept?	❑	❑

Figure 12.1 Air emissions.

Specify the type of waste that is being generated by each business activity, the method of disposal and any relevant licences held and the waste regulation they relate to (Figure 12.3). If you are not sure of the nature or classification of the waste being generated, record it and seek third-party professional advice; the key advisors are listed in Appendix 3.

Soil and groundwater protection

If your company stores chemicals, solvents, paints, diesel, petrol, heating oils, etc., be aware of the consequences of any spillage. The emphasis here is on what the company stores and on where, and how, it is stored. If spillage of hazardous materials occurs, you will need to know what has been spilt, the volume of the spillage, where the spillage will drain to or flow naturally, and where it is likely to end up. A plan of the storage areas and directions of flows in the event of any spillage will be required. Any drainage plan will again be of assistance here.

If some water or land contamination does arise, you will need to identify the method you intend to use for remediation and methods for future protection. If your company currently uses waste contractors, they also need to be identified.

Type of discharge	Activity	Method of discharge	Site legislation/regulation/ authorisation/consent

Who is responsible for managing your wastewater discharge?

	Yes	No
Do you monitor wastewater discharges?	❑	❑
Do you have objectives and targets for wastewater discharges?	❑	❑
Do you have documented procedures for identifying, monitoring and controlling wastewater discharges?	❑	❑
Have you conducted training for wastewater discharge procedures?	❑	❑
Have you audited these procedures?	❑	❑
Is any wastewater discharged directly into controlled waters?	❑	❑
Is stormwater on your site discharged directly into controlled waters?	❑	❑
Is any wastewater treated on your site?	❑	❑
Do you have a site plan illustrating draining plans and discharge points?	❑	❑

If so, where is this plan kept?

Figure 12.2 Water pollution.

 **Hint**

One point to consider – particularly if your company is buying or leasing land or buildings for business use – is the need to check for any previous water or land pollution. Ensure that the selling or letting company remedies the pollution before your company moves in.

Do you produce or handle any of the following?		
	Yes	No
Solid waste	❑	❑
Hazardous waste	❑	❑
Special waste	❑	❑
Clinical waste	❑	❑

Please complete if you produce any solid or hazardous waste			
Waste type	Activity creating this waste	Method of disposal	Site legislation/ regulation/licence

	Yes	No
Do you monitor your waste streams?	❑	❑
Do you have objectives and targets for dealing with waste?	❑	❑
Do you have documented procedures for identifying, monitoring and controlling waste?	❑	❑
Have you conducted training for your waste management procedures?	❑	❑
Have you audited these procedures?	❑	❑
Do you hold your own waste management licence?	❑	❑
Do you hold copies of licences for your waste transporters or disposal sites?	❑	❑
Do you have a register of consignment/transfer notes?	❑	❑
Do you have a site plan of waste generation and storage points?	❑	❑
If so, where is this plan kept?		

Figure 12.3 Solid and hazardous waste.

Noise control

Check noise levels internally and externally. Most modern internally operated machinery generate excessive noise, even if it is only intermittent. External noise is also classified as pollution, and you can be sure that if your company is emitting excessive external noise your neighbours, whoever they may be, will not be slow in telling you and the local authority and Health and Safety Executive (HSE). List those processes or other business activities that emit noise and have them measured by a third party to check the limits. A check on the acceptance threshold levels can be actioned through your company's local environmental health department.

Resource management

This is primarily a look at other resource-saving activities that may have been tested or implemented, e.g. to better manage fuel, energy and water consumption within the company. Take into consideration any waste or energy minimisation surveys that may have been undertaken directly by the company or indirectly by a consultant or other third party. You will be asked whether your company has waste or energy minimisation objectives and targets, whether it has a responsible person monitoring energy use and whether it has any material recycling or reuse initiatives currently in operation. You will also be asked whether your company has considered other uses for the land it occupies.

You may surprise yourself in this section by ticking more 'yes' than 'no' boxes. Most companies tend to focus on various methods of corporate cost savings and most would acquit themselves better under this section than in others.

Other environmental impacts

If you feel that previous sections have not really covered every activity that your company is involved in, this section allows you to consider the environmental impact that things such as odour, dust, vibration and visual impact may have on the environment.

Indirect impacts

Apart from the direct environmental impacts that your business's activities may have on the environment, you will be asked to consider also the indirect impact. The indirect impacts may be difficult to spot initially but can come in the form of, for example, product designs. Better product design may result in energy savings or a greater percentage of the product being recyclable. Better design of some brands of motor car has enabled a greater proportion of the vehicle – up to 85 per cent in some cases – to be recycled or reused.

Another area to consider is the packaging of distributed goods. You will be asked whether your company reuses the packing materials saved from incoming goods and whether your company uses its own transport for delivery. The use of a courier may be a worthwhile option with regard to reducing air emissions and fuel bills. One of the ways towards resolving a proportion of these indirect impacts is to ask your company's suppliers and customers whether there are more environmentally friendly alternatives to existing practices.

Environmental exercise

Following the completion of the self-assessment form, you may be alarmed about the large number of 'no' boxes ticked; do not worry, this is normal and, unless you have undergone previous energy or waste study exercises, you will be fortunate to get many 'yes' ticks. So, what is the point of the exercise? The point is that you have started your company's environmental management assessment, you have little or nothing in place, you are more familiar with the terminology, you are starting to plan how to get more 'yes' ticks and the only way is up. With a minimum of 'yes' boxes ticked, you will be astonished at the rate of improvement at your next assessment stage, which may be only 3 months away.

Completing the preassessment form will give some general points as to what aspects of your company's activities will have the greatest impact upon the environment. This will act as a precursor to the creation of your aspects and impacts analysis (Figure 12.4).

Preassessment report

Before the completion of your company's preassessment visit, it is advisable to discuss possible time-scales for implementation of its chosen environmental management system with the assessor. To demonstrate the content and format of assessment reports, it is again assumed, in the text and examples here, that your company is introducing the ISO 14001 standard.

As previously mentioned, if you and your company are new to the environmental management game, the preassessment report could probably stretch to ten pages. Do not be put off by this. The assessor will assess your company's existing environmental management system, which will be minimal at this point compared with the requirements of the full ISO 14001 standard. It is a strange sensation to go through the preassessment exercise with a view to implementing an environmental management system and to be assessed as if your company had one in place. Although you do not have an EMS, the assessor considers that you

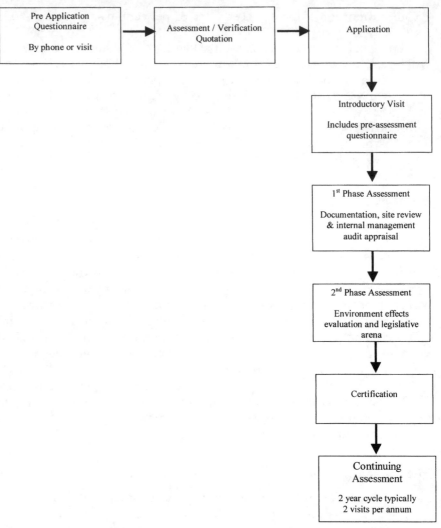

Figure 12.4 Route to certification.

have and audits your company's system effectively against the ISO 14001 standard. It may seem harsh at the time, but you do need to have benchmarks against the standard that your company is pursuing. The assessor is aware of your sensitivities and this is often reflected in the summary of the preassessment report.

The bulk of the report will be an assessment of how your company measures up to the key elements of the standard in areas such as environmental policy, planning, operation, implementation, procedures, auditing and corrective actions. The company will also be subjected to an external site visit. Here, the assessor will identify areas that require improvement. These improvement actions may include the simple removal of old cars, chemical drums and litter, or extend to the more involved issues of land and water contamination and the integrity of existing fuel lines. Remember that the assessor is identifying issues to be remedied, and plenty of time and support is given for this to happen. The assessor is on your side and acts in the strictest confidence.

On completion of the preassessment, a report is produced of the findings. It will have a short executive-type summary that provides an overall impression of the preassessment. It should comprise three main sections: a short introduction, the findings and conclusions. The introduction will state simply that the preassessment has been carried out to record the current status of the existing environmental management system of the company and to provide a realistic time-scale towards final certification of the standard.

The findings section of the report should be very positive in the sense that the assessor identifies everything that has been done within the company that would qualify as environmental improvement. This may include energy surveys, waste recording, recycling of production materials, etc. If your company has initial environmental objectives – one of which may be to implement an environmental standard – and a one-page corporate environmental policy, these will be considered as the beginnings of an environmental management system.

The conclusions of the report will highlight those areas of the company for environmental improvement and refer to those pages in the report identifying specific improvement actions. In this section, the assessor should identify the assessment team that will be assigned to your company. The team will most likely consist of three assessors: a lead assessor, a team assessor and an industry specialist.

Following the report, the assessor may suggest a date for the next assessment, but it is open to negotiation if you feel there is sufficient commitment within the company, and that it is not too large, to progress at a quicker pace. If you want to speed along, 3 months is probably an acceptable period of time before the next assessment date.

You will find that, after every other assessment, an action plan to address points identified will be a requirement. At the preassessment stage, a response to the action points identified is not required. They must still be remedied, but the remedial actions taken will be assessed at the next assessment.

After completion of the preassessment, you, your team and the assessor will talk through the findings of the report and a consensus will be sought on the findings. Do not feel that you have to accept everything in the report. If you have sufficient grounds, or preferably evidence, to argue against a particular finding, do so. Assessors are not infallible and are inclined to give you the benefit of any doubt. Once agreed, however, you will sign the report and accept its findings. You will also sign for the time taken by the assessor and probably acceptance of the fees to be invoiced for that assessment. This report acceptance procedure applies to all assessments and final certification.

Assessments 1 and 2

These two assessments are being discussed jointly to emphasise the fact that it is possible to take more than one assessment at any one time. As noted earlier, you should allow enough time for your company to be ready for each phase. Three months may have elapsed since preassessment before your company is ready for its first assessment. Because the standard's requirements tend to overlap between stages, it is possible and sensible to undertake two phases on the same assessment date. This first phase focuses predominantly on collecting the required paperwork together, i.e. the environmental manual, correspondence folder, documents, logs, etc. Your company will also be subjected to another external assessment of its environmental procedures.

A combined phase 1 and 2 assessment will be a 1-day event with two assessors. One assessor will be the lead assessor and the other will be the industry specialist. The lead assessor concentrates on how your company's procedures relate to the standard's requirements, whereas the industry specialist will concentrate on how your company's procedures relate to its business activities and operations.

The first half of the day will be taken up by the lead assessor going over the preassessment report to ensure that all of the necessary corrective actions have been made. While the lead assessor is doing this, the industry specialist will be prowling around internally and externally looking for operational improvements flagged in the preassessment report. So, in brief, the first half the day is spent checking the previous report and the second half testing whether the paperwork conforms to the standard and that the procedures will work.

Phase 3

This is the final phase, the certification stage of ISO 14001, which requires the full involvement of the assessors' team (three people). The assessment takes 2 days and each member of the team will focus on all aspects of the standard. Everything that has been written, implemented, tested and audited is assessed in these 2 days. It sounds daunting, but if early assessments have progressed well and changes recommended from stage 2 have been achieved – and as long as you have followed the ISO 14001 phase 3 requirements – your company should pass.

It does not matter if you have one or two observations, recommendations or even non-conformities in phase 3. In fact, you can rely on the assessors to find something that is not quite right. The report at the end of the assessment will state the findings, and even if there are still some corrections to be made the certification may be awarded. Before receiving the certificate, however, your company will have to send a report to the lead assessor stating that it has in fact addressed, or intends to address by a certain date, all of the system corrections required.

If the report is acceptable, the assessor will make recommendation to the assessment board for the award of ISO 14001 accreditation and a certificate. You will also be informed that the first system audit will be conducted in 6 months. It is during this test, the first verification audit, that the remedial actions specified in your report will be audited.

Appendix 1   Terminology

It is important to provide definitions of the environmental terms used in the EMS plan. Where possible, however, avoid unnecessary environmental jargon; where jargon is used, ensure that the definition is simple and clear and matches that used in relevant legislation.

Abnormal operating conditions

Operating conditions that occur infrequently or during non-daily activities (e.g. during maintenance, start-up or shut-down procedures).

Business areas

Departments or sites where the environmental management system is to be applied.

Business unit

A building or area that contains activities and/or processes that are under management control. A site can be composed of a number of business managers or project managers, each of whom has defined responsibilities within the environmental management system.

Clause statements

The ISO 14001 standard is broken down into main sections or clauses. Each clause contains a number of statements that must be adhered to and proven if certification is to be obtained.

Company

An organisation, corporation, firm, enterprise, site, local authority or institution, whether in the public or private sector, that has its own business functions and administration.

Continual improvement

The on-going process of enhancing the EMS to achieve improvements in overall environmental performance in line with the corporate environmental policy.

Corporate environmental policy

A statement by the company of its intentions and principles in relation to its environmental performance that provides a framework for action and for the setting of its environmental objectives and targets.

EMS certification of verification audit

An audit conducted by external assessors from a registration body (e.g. British Standards Institution) against the requirements of the standard. It is often conducted in four stages:

- Preliminary audit. A check of the system by a certification body – not compulsory but recommended.
- Stage 1 assessment. Documentation review against the requirements of the standard. This includes creating an audit programme for the next stage.
- Stage 2 assessment. This focuses on the implementation and continuing application of the environmental management system. Certification will or will not be awarded at this stage.
- Surveillance visits. Routine programmed visits at approximately 6-monthly intervals to assess improvements in environmental performance.

Environment

The whole of the surroundings in which the company operates, including air, water, land, natural resources, flora, fauna, humans and their interrelations.

Environmental aspect

That element or aspect of the company's activities, products or services which interacts with the environment.

Environmental correspondence folder

A log of all internal and external correspondence that relates to the operation of the environmental management system.

Environmental impact

Any change to the environment, whether adverse or beneficial, wholly or partially resulting from the company's activities, products or services.

Environmental impact rating

A scoring system used for assessing the significance of any identified environmental aspects associated with a company's processes and activities.

Environmental management manual

A document describing the overall EMS and making reference to the procedures for implementing the organisation's environmental management policy.

Environmental management procedures

An information sheet, or document, that provides guidance concerning the management of control procedures within the environmental management system. The procedure can either stand alone or be incorporated into an existing procedure. Environmental management procedures are frequently incorporated into the environmental management manual.

Environmental management review

A formal evaluation by senior management of the status and adequacy of a company's environmental policy, systems and procedures in relation to environmental issues, regulations and changing circumstances.

Environmental management system (EMS)

The part of the overall management system that includes organisational structure, planning activities, responsibilities, practices, procedures, processes and resources for developing, implementing, achieving, reviewing and maintaining the environmental policy.

Environmental management system audit

A systematic and documented verification process of obtaining and evaluating evidence objectively to determine whether the company's EMS conforms to the relevant audit criteria set by the company, and for communication of the results of this process to management.

Environmental objective

The overall environmental goal, arising from the environmental policy, that the company has set itself, and which is quantified where possible.

Environmental operating procedures

An information sheet, or document, that provides guidance concerning the operation of tasks within the environmental management system. The procedure can either stand alone or be incorporated into an existing procedure. ISO 14001 procedures are frequently incorporated into the environmental management manual.

Environmental performance

Measurable results of the EMS that are related to the company's control of its environmental aspects, based on its environmental policy, objectives and targets.

Environmental significance

Assessment criteria used to determine the significance of environmental impact. The significance rating is based on the environment impact of the aspect while taking into account the site setting and sensitivity.

Environmental target

A detailed performance requirement – quantified where practicable – applicable to the company, or parts thereof, that arises from the environmental objectives and that needs to be set and met in order to achieve those objectives.

Initial environmental review

A process by which a company can establish its current position with regard to internal and external environmental issues.

Interested party

An individual or group concerned with or affected by the environmental performance of the company.

Normal operating conditions

Operating conditions that occur regularly or as a result of normal daily business activities.

Prevention of pollution

The use of processes or practices, materials or products that avoid, reduce or control pollution, which may include recycling, treatment, process changes, control mechanisms, efficient use of resources and material substitution.

Responsible person

A person who has clearly defined and documented responsibilities for ensuring that all or part of the environmental management system is established and maintained.

Significant environmental aspect

A significant environmental aspect is an environmental aspect that has or can have a significant environmental impact as a result of a high rating.

Site/department

A site, or department, that has clearly defined and documented environmental responsibilities (including regulatory authority communication, contractor control, bund inspections, water sampling, etc.).

Top management

Top management may consist of an individual or group of individuals with policy decision-making responsibility for the company.

Appendix 2
A guide to environmental legislation and regulation

Depending on the level of your business's activity and the industrial sector it is in, some environmental legislation will apply and some will not. The legislation guide includes existing legislation. Proposed legislation, such as the waste electrical and electronic equipment (WEEE) and waste paper and packaging recycling legislation and the climate change levy, have been omitted. These pieces of environmental legislation are just three of several that are currently under consultation. Eventually, they will become law, but this may take another 2–3 years, by which time the form and content may have changed considerably.

The ISO 14001 standard requires that a company monitors environmental legislation and lists, in a register, any that becomes law and has an impact on it. It is prudent to monitor draft legislation as well to determine whether it could have an impact upon your company. Those companies that have, for example, a basic idea of how the WEEE regulations may affect them have the opportunity to take early remedial action or to make comment and lobby for change. By taking this early action, they will gain advantage over their competitors and reduce their exposure to future environmental liability.

The following is a list of environmental legislation. It is not meant to be exhaustive, but merely to provide companies with an insight into the variety of pieces of legislation that may affect them.

A

Air Quality Standards Regulations 1989 SI 317
Air Quality Standards (Amendment) Regulations 1995 SI 3146

B

Bathing Waters (Classification) Regulations 1991 SI 1597
Bathing Waters (Classification) (Scotland) Regulations 1991 SI 1609
Batteries and Accumulators (Containing Dangerous Substances) Regulations 1994 SI 232

C

Carriage of Dangerous Goods (Classification, Packaging and Labelling) and Use of Transportable Pressure Receptacles Regulations 1996 SI 2092

Carriage of Dangerous Goods by Rail Regulations 1996 SI 2089

Carriage of Dangerous Goods by Road Regulations 1996 SI 2095

Carriage of Dangerous Goods by Road (Driver Training) Regulations 1996 SI 2094

Carriage of Explosives by Road Regulations 1996 SI 2093

Chemicals (Hazard Information and Packaging) Regulations 1994 SI 3247 7.6

Chemicals (Hazard Information and Packaging for Supply) Regulations 1996 SI 1092

Classification and Labelling of Explosives Regulations 1982 SI 1140

Clean Air (Arrestment Plant) (Exemption) Regulation 1969 SI 1262

Clean Air (Arrestment Plant) (Exemption) (Scotland) Regulations 1969 SI 1386

Clean Air (Emission of Dark Smoke) (Exemption) Regulations 1969 SI 1263

Clean Air (Emission of Dark Smoke) (Exemption) (Scotland) Regulations 1969 SI 1389

Clean Air (Emission of Grit and Dust from Furnaces) 1971 SI 162

Clean Air (Height of Chimneys) (Exemption) Regulations 1969 SI 411

Clean Air (Height of Chimneys) (Exemption) (Scotland) Regulations 1969 SI 465

Control of Pollution (Exemption of Certain Discharges from Control) Order 1983 SI 11821986 SI 1623

Control of Pollution (Exemption of Certain Discharges) (Variation) Order 1987 SI 1782

Control of Pollution (Exemption of Certain Discharges from Control) (Scotland) (Variation) Order 1993 SI 1154

Control of Pollution (Registers) (Scotland) Regulations 1993 SI 1155

Control of Pollution (Silage, Slurry and Agricultural Fuel Oil) Regulations 1991 SI 324

Control of Pollution (Silage, Slurry and Agricultural Fuel Oil) (Amendment) Regulations 1996 SI 2044 1997 SI 547

Control of Pollution (Silage, Slurry and Agricultural Fuel Oil) (Scotland) Regulations 1991 SI 346

Control of Pollution (Supply and Use of Injurious Substances) Regulations 1986 SI 902

Control of Pollution Act 1974 Hazardous Substances Noise

Control of Substances Hazardous to Health Regulations 1994 SI 3246

Control of Substances Hazardous to Health (Amendment) Regulations 1996 SI 3138 1997 SI 11

Controlled Waste (Registration of Carriers and Seizure of Vehicles) Regulations 1991 SI 1624

Controlled Waste Amendment Regulations 1993 SI 566

Controlled Waste Regulations 1992 SI 588
Controlled Water (Lakes and Ponds) Order 1989 SI 1149
Controlled Waters (Lochs and Ponds) (Scotland) Order 1990 SI 120

D

Dangerous Substances (Notification and Marking of Sites) Regulations 1990 SI 304
Dark Smoke (Permitted Periods) (Scotland) Regulations 1958 SI 1993
Dark Smoke (Permitted Periods) (Vessels) Regulations 1958 SI 878
Dark Smoke (Permitted Periods) (Vessels) (Scotland) (Regulations) 1958 SI 1934
Dark Smoke (Permitted Periods) Regulations 1958 SI 498
Deposit of Poisonous Wastes Act 1972
Designation of Structure Plan Areas (Scotland) Order 1996

E

EC Directives
   Air Pollution Emissions from Large Combustion
   Hazardous Substances Continued Marketing and Use of Dangerous Substances 76/769/EEC 3rd Amendment 82/828/EEC, 6th amendment 85/467/EEC, 8th amendment 89/677/EEC, 9th amendment 91/173/EEC, 10th amendment 89/338/EEC, 11th amendment 91/339/EEC
   Laws Relating to Classification and Labelling of Dangerous Substances 88/379/EEC, 7th amendment 92/32/EEC
   Major Accident Hazards of Certain Industrial Activities ('Seveso') 82/501/EEC
   Placing of Plant Protection Products on the Market 91/414/EEC
   Specific Information on Dangerous Preparations (safety data sheets) 91/155/EEC
   Transport of Dangerous Goods by Road and Rail 94/55/EEC
   Use of Certain Active Substances in Pesticides 79/117/EEC
   Waste Management Packaging and Packaging Waste 94/62/EEC
   Urban Waste Treatment 91/271/EEC
   Waste Framework Directive 75/442/EEC, amendment 91/157/EEC
EC Regulations Allowing a Community Eco-Management and Audit Scheme EEC/1836/93
   Regulation on a Community Eco-Label Award Scheme EEC/880/92
   Regulation on Substances that Deplete the Ozone Layer EEC/3093/95
   Regulation on the Supervision and Control of Wastes in the European Community EEC/259/93
   Regulation on Evaluation and Control of Existing Substances EEC/793/93

Electricity and Pipe-line Works (Assessment of Environmental Effects) Regulations 1990 SI 442

Environment Act 1995 Air Pollution Producer Responsibility

Environmental Assessment (Aforestation) Regulations 1988 SI 1207

Environmental Assessment (Salmon Farming in Marine Waters) Regulations 1988

Environmental Assessment (Scotland) Regulations 1988 SI 1221

Environmental Assessment (Scotland) (Amendment) Regulations 1994 SI 2012

Environment Protection (Applications, Appeals and Registers) Regulations 1991 SI 507

Environmental Protection (Applications, Appeals and Registers) (Amendment) Regulations 1996 SI 667, 1996 (NO2) SI 979

Environmental Protection (Authorisation of Processes) (Determination Periods) Order 1991 SI 513

Environmental Protection (Authorisation of Processes) (Determination Periods) (Amendment) Order SI 2847

Environmental Protection (Controls on Injurious Substances) Regulations 1992 SI 131 1993 SI 1

Environmental Protection (Controls on Injurious substances) (No. 2) Regulations 1992 SI 1583 1993 SI 1643

Environmental Protection (Controls on Substances that Deplete the Ozone Layer) 1996 SI 506

Environmental Protection (Duty of Care) Regulations 1991 2839

Environmental Protection (Prescribed Processes and Substances) Regulations 1991 SI 472

Environmental Protection (Prescribed Processes and Substances) (Amendment) Regulations 1991 SI 836 1992 SI 614 1993 SI 1749 1993 (No. 2) SI 2405 1994 SI 127 1994 (No. 2) SI 1329 1995 SI 3247

Environmental Protection (Prescribed Process and Substances) (Amendment) (Petrol Vapour Recovery) 1996 SI 2678

Environmental Protection Act 1990 Air Pollution Development Control Effluent and Water Hazardous Substances IPC Noise Nuisance Waste Management

Explosives Act 1875

Explosives Act 1923

F

Finance Act 1996

Food and Environment Protection Act 1985

H

Harbour Works (Assessment of Environmental Effects) Regulations 1996 SI 1946

Planning (Hazardous Substances) Act 1990
Planning (Hazardous Substances) (Scotland) Act 1997
Planning (Hazardous Substances) Regulations 1992 SI 656
Planning (Listed Building and Conservation Areas) Act 1990
Planning and Compensation Act 1991
Plant Protection Products Regulations 1995 SI 887
Private Water Supplies Regulations 1991 SI 1837
Private Water Supplies (Scotland) Regulations 1992 SI 575
Producer Responsibility Obligations (Packaging Waste) Regulations 1997
    SI 648
Protection of Badgers Act 1992
Protection of Birds Act 1954

R

Radioactive Material (Road Transport) Act 1991
Radioactive Material (Road Transport) (Great Britain) Regulations 1996 SI
    1350
Radioactive Substances Act 1993

S

Sewerage (Scotland) Act 1968
Sludge (Use in Agriculture) Regulations 1989 SI 1623
Sludge (Use in Agriculture) (Amendment) Regulations 1990 SI 880
Special Waste Regulations 1996 SI 972
Special Waste (Amendment) Regulations 1996 SI 2019 1997 SI 251
Special Waste (Scotland) Regulations 1997 SI 257
Statutory Nuisance (Appeals) Regulations 1995 SI 2644
Statutory Nuisance (Appeals) (Scotland) Regulations 1996 SI 1076
Surface Waters (Abstraction for Drinking Water) Classification Regulations
    1996 SI 3001
Surface Waters (Abstraction for Drinking Water) (Classification) (Scotland)
    Regulations 1990 SI 121
Surface Waters (Classification) Regulations 1989 SI 1148
Surface Waters (Classification) (Scotland) Regulations 1990 SI 121
Surface Waters (Dangerous Substances) (Classification) Regulations 1989
    SI 2286 1992 SI 337
Surface Waters (River Ecosystems) (Classification) Regulations 1994 SI 1057

T

Town and Country Planning Act 1990
Town and Country Planning (Assessment of Environment Effects)
    (Amendment) Regulations 1988 SI 1199 1990 SI 1218 1992 SI 1494 1994
    SI 677

Town and Country Planning (Development Contrary to Development Plans) (Scotland) Direction 1996

Town and Country Planning (Development Plans and Consultation) Direction 1992

Town and Country Planning (Environmental Assessment and Permitted Development) Regulations 1995 SI 417

Town and Country Planning (Environmental Assessment and Un-authorised Development) Regulations 1995 SI 2258

Town and Country Planning (General Development Procedure) Order 1995 SI 419

Town and Country Planning (General Development Procedure) (Scotland) Order 1992 SI 224

Town and Country Planning (General Development Procedure) (Scotland) Amendment Order 1997 SI 749

Town and Country Planning (General Permitted Development) Order 1995 SI 418

Town and Country Planning (General Permitted Development) (Scotland) Order 1992 SI 223

Town and Country Planning (Hazardous Substances) (Scotland) Regulations 1993 SI 323

Town and Country Planning (Listed Building and Conservation Areas) (Scotland) Regulations 1987 SI 1529

Town and Country Planning (Notification of Applications) (Scotland) Direction 1997

Town and Country Planning (Scotland) Act 1972

Town and Country Planning (Use Classes) Order 1987 SI 764

Town and Country Planning (Use Classes) (Scotland) Order 1989 SI 147

Town and Country Planning (Use Classes) (Scotland) (Amendment) Order 1993 SI 1038

Trade Effluent (Asbestos) (Scotland) Regulations 1993 SI 1446

Trade Effluent (Prescribed Processes and Substances) Regulations 1989 SI 1156 1992 SI 339

Trade Effluent (Prescribed Processes and Substances) (Amendment) Regulations 1990 SI 1629

Transfrontier Shipment of Waste Regulations 1994 SI 1137

U

Urban Waste Water Treatment (England and Wales) Regulations 1994 SI 2841

Urban Waste Water Treatment (Scotland) Regulations 1994 SI 2842

W

Waste Management Licensing Regulations 1994 SI 1056

Waste Management Licensing (Amendment) Regulations 1995 SI 288 1995 (No. 2) SI 1950 1996 SI 1279

Water (Prevention of Pollution) (Code of Practice) Order 1991 SI 2285

Water (Prevention of Pollution) (Code of Practice) (Scotland) Order 1992 SI 395

Water Act 1989

Water (Scotland) Act 1980

Water Industry Act 1991

Water Resources (Licences) Regulations 1965 SI 534 1989 SI 336

Water Resources (Licences) Amendment Regulations 1965 SI 2082

Water Resources Act 1991

Water Supply (Water Quality) Regulations 1989 SI 1147

Water Supply (Water Quality) (Amendment) Regulations 1991 SI 1384

Water Supply (Water Quality) (Scotland) (Amendment) Regulations 1990 SI 119 1991 SI 1333

Wildlife and Countryside Act 1981

Appendix 3
Environmental management support services

Divided into two main sections, this appendix provides many useful UK and European contact points for sources of environmental information. The first part provides sources of environmental information for those companies operating in the UK and Europe. The second part identifies those accreditation agencies that exist and gives their office locations around the world.

Environment Agency

Enquiry line – Tel: (0645) 333111 (connects directly to your company's local office)
Pollution reporting – Tel: (0800) 807060
Website – www.environment-agency.gov.uk
E-mail – enquiries@environment-agency.gov.uk

Bristol HQ

Tel: (01454) 624400; Fax: (01454) 624409

Anglian region

HQ – Tel: (01733) 371811; Fax: (01733) 231840

Central area
Tel: (01480) 414581; Fax: (01480) 413381

Eastern area
Tel: (01473) 727712; Fax: (01473) 724205

Northern area
Tel: (01522) 513100; Fax: (01522) 512927

Midlands region

HQ – Tel: (0121) 711 2324; Fax: (0121) 711 5824

Lower Severn area
Tel: (01684) 850951; Fax: (01684) 293599

Upper Severn area
Tel: (01743) 272828; Fax: (01743) 272139

Lower Trent area
Tel: (0115) 945 5722; Fax: (0115) 981 7743

Upper Trent area
Tel: (01543) 444141; Fax: (01543) 444161

North-east region

HQ – Tel: (0113) 244 0191; Fax: (0113) 246 1889

Dales area
Tel: (01904) 692296; Fax: (01904) 693748

Northumbria area
Tel: (0191) 203 4000; Fax: (0191) 203 4004

Ridings area
Tel: (0113) 244 0191; Fax: (0113) 231 2116

North-west region

HQ – Tel: (01925) 653999; Fax: (01925) 415961

Central area
Tel: (01772) 39882; Fax: (01772) 627730

North area
Tel: (01228) 25151; Fax: (01228) 49734

South area
Tel: (0161) 973 2237 ;Fax: (0161) 973 4601

Southern region

HQ – Tel: (01903) 832000; Fax: (01903) 821832

Hampshire area
Tel: (01962) 713267; Fax: (01962) 841573

Isle of Wight
Tel: (01983) 822986; Fax: (01983) 822985

Kent area
Tel: (01732) 875587; Fax: (01732) 875057

Sussex area
Tel: (01903) 215835; Fax: (01903) 215884

South-west region

HQ – Tel: (01392) 444000; Fax: (01392) 444238

Cornwall area
Tel: (01208) 78301; Fax: (01208) 78321

Devon area
Tel: (01392) 444000; Fax: (01392) 444238

North Wessex area
Tel: (01278) 457333; Fax: (01278) 452985

South Wessex area
Tel: (01258) 456080; Fax: (01258) 455998

Thames region

HQ – Tel: (01189) 535000; Fax: (01734) 500388

North-east area
Tel: (01992) 635566; Fax: (01992) 645468

South-east area
Tel: (01932) 789833; Fax: (01932) 786463

West area
Tel: (01491) 535000; Fax: (01734) 535900

Welsh region

HQ – Tel: (01222) 770088; Fax: (01222) 798555

Northern area
Tel: (01248) 670770; Fax: (01248) 670561

South-east area
Tel: (01222) 770088; Fax: (01222) 798555

South-west area
Tel: (01437) 760081; Fax: (01437) 760881

Environment Agency national centres of expertise

Ecotoxicology and Hazardous Substances

Tel: (01491) 832801; Fax: (01491) 834703

Environmental Data and Surveillance

Tel: (01278) 457333; Fax: (01225) 469939

National Centre for Compliance Assessment

Tel: (01524) 842704; Fax: (01524) 842709

National Coarse Fisheries Centre

Tel: (01562) 863887; Fax: (01562) 69477

National Groundwater and Contaminated Land Centre

Tel: (0121) 7112324; Fax: (0121) 7115925

National Water Demand Management Centre

Tel: (01903) 832000; Fax: (01903) 832274
E-mail: wdmc@dial.pipex.com

Risk Analysis and Options Appraisal

Tel: (020) 7664 6811; Fax: (020) 7664 6911

Salmon and Trout Fisheries Science

Tel: (01222) 770088; Fax: (01222) 798383

Scottish Environment Protection Agency
Head office – Tel: (01786) 457700
Pollution reporting – Tel: (0345) 737271
Website: www.sepa.org.uk
E-mail: info@sepa.org.uk

East region

HQ – Tel: (0131) 449 7296; Fax: (0131) 449 7277

Arbroath office
Tel: (01241) 874370; Fax: (01241) 430695

Galashiels office
Tel: (01896) 754797; Fax: (01896) 754412

Glenrothes office (air/waste)
Tel: (01592) 645565; Fax: (01592) 645567

Glenrothes office (water)
Tel: (01592) 759361; Fax: (01592) 759446

Perth office
Tel: (01738) 627989; Fax: (01738) 630997

Stirling office
Tel: (01786) 461407; Fax: (01786) 461425

North region

HQ – Tel: (01349) 862021; Fax: (01349) 863987

Aberdeen office
Tel: (01224) 248338; Fax: (01224) 248591

Elgin office
Tel: (01343) 547663; Fax: (01343) 540884

Fort William office
Tel: (01397) 704426; Fax: (01397) 705404

Fraserburgh office
Tel: (01346) 510502; Fax: (01346) 515444

Orkney office
Tel: (01856) 871080; Fax: (01856) 871090

Shetland office
Tel: (01595) 696926; Fax: (01595) 696946

Thurso office
Tel: (01847) 894422; Fax: (01847) 893365

Western Isles office
Tel: (01851) 706477; Fax: (01851) 703510

West region

HQ – Tel: (01355) 238181; Fax: (01355) 264323

Ayr office
Tel: (01292) 264047; Fax: (01292) 611130

Dumfries office
Tel: (01387) 720502; Fax: (01387) 721154

Lochgilphead office
Tel: (01546) 602876; Fax: (01546) 602337

Newton Stewart office
Tel: (01671) 402618; Fax: (01671) 404121

Northern Ireland Environment and Heritage Service

Pollution reporting: (01232) 757414
Website: www.nics.gov.uk
E-mail: EHS@nics.gov.uk

Conservation Designations and Protection

Tel: (01232) 546612

Conservation Science

Tel: (01232) 546592

Countryside and Coastal Management

Tel: (01232) 546555

Customer Services

Tel: (01232) 546533

Drinking Water Inspectorate

Tel: (01232) 254862

Environmental Quality Unit

Tel: (01232) 254816

Industrial Air Pollution and Radiochemical Inspectorate

Tel: (01232) 254709

Information (leaflets/publications)

Tel: (01232) 546528

Regional Operations

Tel: (01232) 546521

Waste Management Inspectorate

Tel: (01232) 254815

Water Quality Unit
Tel: (01232) 254757

Other government and regulatory contacts

Commission of the European Communities

England
Jean Monnet House
8 Storey's Gate
London SW1P 3AT
Tel: (020) 7973 1992

Northern Ireland
Windsor House
9/15 Bedford Street
Belfast BT2 7EG
Tel: (01232) 240708

Scotland
9 Alva Street
Edinburgh EH2 4PH
Tel: (0131) 2252058

Wales
4 Cathedral Road
Cardiff CF1 9SG
Tel: (01222) 371631

Department of the Environment, Transport and the Regions (DETR)

Eland House
Bressenden Place
London SW1E 5DU
Tel: (020) 7890 3333
Publications order line – Tel: (020) 8691 9191
Website: www.open.gov.uk

Department of Trade and Industry

1 Victoria Street
London SW1H 0ET
Tel: (020) 7215 5000
Publications order line – Tel: 0870 1502 500
Website: www.open.gov.uk

Drinking Water Inspectorate

Romney House
43 Marsham Street
London SW1P 3PY
Tel: (020) 7276 8296

Energy Efficiency Office

See also Government Offices for regional contacts in England

England
Environmental and Energy Management Directorate
Ashdown House
123 Victoria Street
London SW1E 6DE
Tel: (020) 7890 6655; Fax: (020) 7890 6689

Scotland
Electronics (Intern) Scottish Office
Victoria Quay
Edinburgh EH6 6QQ
Tel: (0131) 244 7130; Fax: (0131) 244 7145

Wales
Cathays Park
Cardiff CF1 1NQ
Tel: (01222) 823126; Fax: (01222) 823661

European Environment Agency

Kongens Nytorv 6
DK-1050 Copenhagen K
Denmark
Tel: (+45) 33 36 71 00
Fax: (+45) 33 36 71 99
E-mail: eea@eea.dk
Website: www.eea.dk
UK national focal point at the DETR – Tel: (020) 7276 8947

Government offices for the regions

Eastern unit
7 Enterprise House
Vision Park
Chivers Way
Cambridge CB4 4ZR
Tel: (01234) 796332; Fax: (01234) 796252

East Midlands
Belgrave Centre
Stanley Place
Talbot Street
Nottingham NG1 5GG
Tel: (0115) 971 9971; Fax: (0115) 971 2404

London
Riverwalk House
157–161 Millbank
London SW1P 4RR
Tel: (020) 7217 3098; Fax: (020) 7217 3465

Merseyside
Cunard Building
Pier Head
Liverpool L1 1QB
Tel: (0151) 224 6300; Fax: (0151) 224 6471

North-east
Stangate House
2 Groat Market
Newcastle upon Tyne NE1 1YN
Tel: (0191) 201 3300; Fax: (0191) 202 3806

North-west
2010 Sunley Tower
Piccadilly Plaza
Manchester M1 4BE
Tel: (0161) 952 4000; Fax: (0161) 952 4004

South-east
Bridge House
1 Walnut Tree Close
Guildford GU1 4GA
Tel: (01483) 882255; Fax: (01483) 882269

South-west
Fourth Floor
The Pithay
Bristol BS1 2PB
Tel: (0117) 900 1700 Fax: (0117) 900 1901

West Midlands
77 Paradise Circus
Queensway
Birmingham B1 2DT
Tel: (0121) 212 5000; Fax: (0121) 212 1010

Yorkshire and Humberside
Seventh Floor
East Wing
City House
Leeds LS1 4US
Tel: (0113) 280 0600; Fax: (0113) 244 9313

HSE Information Centre

Broad Lane
Sheffield S3 7HQ
Infoline – Tel: (0541) 545500
HSE Book – Tel: (01787) 881165
Fax: (01787) 313995
Website: www.open.gov.uk

Ministry of Agriculture, Fisheries and Food

Whitehall Place
London SW1A 2HH
Tel: (0645) 335577
Website: www.open.gov.uk

Office of Electricity Regulation (OFFER)

Hagley House
Hagley Road
Edgebaston
Birmingham B16 8QG
Tel: (0121) 456 2100; Fax: (0121) 456 4664
Website: www.open.gov.uk

Office of Gas Supply (OFGAS)

Stockley House
130 Wilton Road
London SW1V 1LQ
Tel: (020) 7828 0898; Fax: (020) 7630 8164

Office of Water Services (OFWAT)

Centre City Tower
7 Hill Street
Birmingham B5 4UA
Tel: (0121) 625 1300; Fax: (0121) 625 1422

Minicom: (0121) 625 1422
Website: www.opem.gov.uk

Planning Inspectorate

Tollgate House
Houlton Street
Bristol BS2 9DJ
Tel: (0117) 987 8000; Fax: (0117) 987 8406

Scottish Office

Pentland House
Robbs Loan
Edinburgh EH14 1TY
Tel: (0131) 556 8400
Website: www.open.gov.uk

Water authorities (Scotland)

East of Scotland Water
Pentland Gait
597 Calder Road
Edinburgh EH11 4HJ
Tel: (0131) 453 7500; Fax: (0131) 453 7554

North of Scotland Water
Cairngorm House
Beechwood Business Park
Inverness IV2 3ED
Tel: (01463) 245400; Fax: (01463) 245405

West of Scotland Water
419 Balmore Road
Glasgow G22 6NU
Tel: (0141) 355 3555; Fax: (0141) 355 5480

Water and sewerage companies (England and Wales)

Anglian Water Services
Anglian House

Ambury Road
Huntingdon PE18 6NX
Tel: (01480) 443000; Fax: (01480) 443115
Website: www.anglianwater.co.uk

Dwr Cymru Welsh Water
Plas-y-Ffynonn
Cambrian Way
Brecon
Powys LD3 7HP
Tel: (01874) 623181; Fax: (01874) 625620

Northumbrian Water
Abbey Road
Pity Me
County Durham DH1 5FJ
Tel: (0191) 383 2222; Fax: (0191) 383 1209

North West Water
Dawson House
Liverpool Road
Warrington WA5 3LW
Tel: (01925) 234000; Fax: (01925) 233360

Severn Trent Water
2297 Coventry Road
Sheldon
Birmingham B26 3PU
Tel: (0121) 722 4000; Fax: (0121) 722 4800
Website: www.severntrent.co.uk

Southern Water
Southern House
Yeoman Road
Worthing BN13 3NX
Tel: (01903) 264444; Fax: (01903) 262100

South West Water
Peninsula House
Ryon Lane
Exeter EX2 7HR
Tel: (01392) 446688; Fax: (01392) 434966
Website: www.sww.co.uk

Thames Water
Nugent House
Vastern Road
Reading RG1 8DB
Tel: (0118) 959 3371; Fax: (0118) 959 3300

Wessex Water
Wessex House
Passage Street
Bristol BS2 0JQ
Tel: (0117) 929 0611; Fax: (0117) 929 3137

Yorkshire Water Services
West Riding House
67 Albion Street
Leeds LS1 5AA
Tel: (0113) 244 8201; Fax: (0113) 244 3071

Welsh Office

Crown Building
Cathays Park
Cardiff CF1 3NQ
Website: www.open.gov.uk

European accreditation agencies

Country	Body	Telephone no.
Denmark	DANAK	(+45) 35 868686
Finland	FINAS	(+358) 0 61671
France	COFRAC	(+33) 1 44 688220
Germany	Bundesumwelt Ministerium	(+49) 228 052256
Greece	Ministry for the Environment	(+30) 1 6465762
Ireland	National Accreditation Board	(+353) 1 6686557
Italy	Ministerio deli Ambiente	(+39) 6 7029215
Luxembourg	Ministerie de l'Environnement	(+352) 4786816
Norway	Norwegian Accreditation	(+47) 22200226
Portugal	Direccao-Geral Ambiente)	(+351) 1 8471022
Spain	MOPTMA	(+34) 1 5977470
Sweden	SWEDAC	(+46) 33 17 7700
The Netherlands	Road voor Certificate	(+31) 343 812604
UK	UKAS	(+44) 20 7233 7111

Registration companies

The companies listed below will be able to assist with registration for ISO 14001, EMAS or BS 7750. In some cases, these companies may be able to offer registration for all three standards.

Austria

BVQI

Marokkanergasse 22/3
A-1030 Vienna
Austria
Tel: (+43) 1 713 15 68
Fax: (+43) 1 712 54 35
Telex: 136575 BVVI A

Belgium

S. A. Bureau Veritas Quality International (Belgium) N.V.
Placc Bara 26 Bte 17/19
B-1070 Brussels
Belgium
Tel: (+32) 2 520 2090
Fax: (+32) 2 520 2030

Canada

SGS International Certification Services Canada
90 Gough Road
Unit 4
Markham
Ontario L3R 5V5
Canada
Tel: (+1) 416 479 1160
Fax: (+1) 416 479 9452

Finland

Finnish Standards Association SFS
Maistraatinportti 2
FIN-00240 Helsinki
Finland
Tel: (+358) 0 149 9331
Fax: (+358) 0 1499 3323
E-mail: sfs@sfs.fi
Link to www page: http://www.sfs.fi/

France

BVQI France
Immeuble Apollo
10 Rue Jacques Daguerre
92565 Rueil-Malmaison, Cedex
France
Tel: (+33) 1 47 14 43 30
Fax: (+33) 1 47 14 43 25

Germany

BVQI
Sachsenfeld 4
Haus 5
Hansa Carree
Postfach 100940
D20006 Hamburg 1
Germany
Tel: (+49) 40 23 62 5120
Fax: (+49) 40 23 62 5100
Telex: 2165767 BUVHD

BVQI
Huyssennallee 5
D-45128 Essen
Germany
Tel: (+49) 201 810 7614
Fax: (+49) 201 810 7620
Telex: 17 201486 BUVE D

Ireland

NSAI (National Standards Authority of Ireland)
Glasnevin
Dublin 9
Ireland
Tel: (+353) 1 8370101
Fax: (+353) 1 8369821

Sweden

BVQI Sweden
Stora Badhusgatan 20
S-411 21 Gothenburg
Sweden
Tel: (+46) 31 17 14 15
Fax: (+46) 31 13 39 73

USA

BSI Quality Assurance
8000 Towers Crescent Drive
Suite 1350
Vienna VA 22182
USA
Tel: (+1) 703 760 7828
Fax: (+1) 703 761 2770

Bureau Veritas QI (NA), Inc. (Central Office)
509 North Main Street
Jamestown NY 14701
USA
Tel: (+1) 716 484 9002 or (+1) 800 937 9311
Fax: (+1) 716 484 9003

Det Norske Veritas Industry, Inc.
16340 Park Ten Place
Suite 100
Houston TX 77084
USA
Tel: (+1) 713 579 9003
Fax: (+1) 713 579 1360

International Approval Services
8501 E. Pleasant Valley Road
Cleveland OH 44131
USA
Tel: (+1) 216 524 4990
Fax: (+1) 216 642 3463
E-mail: ias@apk.net

Lloyds Register QA
33–41 Newark Street
Riverview Historical Plaza
Hoboken NJ 07030
USA
Tel: (+1) 201 963 1111
Fax: (+1) 201 963 3299

National Quality Assurance, Inc.
1146 Massachusetts Avenue
Boxborough MA 01719
USA
Tel: (+1) 508 635 9256
TollFree: (+1) 800 649 5289
Fax: (+1) 508 266 1073
E-mail: nqa@interramp.com

SGS International Certification Services
1415 Park Avenue
Hoboken NJ 07030
USA
Tel: (+1) 800 777 8378
Fax: (+1) 201 792 2558

UK

Aspects Certification Services Limited
864 Birchwood Boulevard
Birchwood
Warrington
Cheshire WA3 7QZ
Tel: (01925) 852851
Fax: (01925) 852857

BVQI Ltd
London Central Office
70 Borough High Street
London SE1 1XF
Tel: (020) 7378 8113
Fax: (020) 7378 8014

BVQI Ltd
14 Challenge House
Sherwood Drive
Bletchley
Milton Keynes MK3 6DP
Tel: (01224) 212838
Fax: (01224) 210924

BVQI Ltd
Suite G1
Norton Centre
Poynernook Road
Aberdeen AB1 2RN
Tel: (01224) 212838
Fax: (01224) 210924

BVQI Ltd
223 Wolverhampton Street
Dudley
West Midlands DY1 1EF
Tel: (01384) 459546
Fax: (01384) 238741
Telex: 886201 BVLONDG

BSI Quality Assurance
PO Box 375
Milton Keynes MK14 6LL
Tel: (01908) 220908
Fax: (01908) 220671

Det Norsk Veritas QA Ltd
Palace House
3 Cathedral Street
London SE1 9DE
Tel: (020) 7357 6080
Fax: (020) 7357 6048

Lloyds Register Quality Assurance
Norfolk House
Wellesley Road
Croydon CR9 2DT
Tel: (020) 8688 6882
Fax: (020) 8681 8146

Professional Environmental Caring and Services QA
Resource House
144 High Street
Rayleigh
Essex SS6 7RU
Tel: (01268) 770135
Fax: (01268) 774436

SGS Yarsley
Trowers Way
Redhill
Surrey RH1 2JN
Tel: (01737) 768445
Fax: (01737) 761229

SGS Yarsley
Unit 7 Ensign Business Centre
Westwood Business Park
Coventry CV4 8JA
Tel: (01203) 694818
Fax: (01203) 470063

TRADA Certification Ltd
Stocking Lane
Hughenden Valley
High Wycombe
Bucks HP14 4NR
Tel: (01494) 565484
Fax: (01494) 565487

Bibliography

British Standards Institution (1996) *Implementation of ISO 14001. Environmental Management Systems – Specification with Guidance for Use*, London: British Standards Institution.

Charter, M. (1992) *Greener Marketing*, Sheffield: Greenleaf Publishing.

Confederation of British Industry (published monthly) *Newsletter*, London: Confederation of British Industry (CBI) Publication.

Elkington, J., Burke, T. and Hailes, J. (1988) *Green Pages: The Business of Saving the World*, London: Routledge.

Environmental Data Services (ENDS) (2000) *The ENDS Report*, London: Environmental Data Services.

Goodland, R., Daly, H., El Sarafy, S. and Droste, B.V. (1992) *Environmentally Sustainable Economic Development: Building on Brundtland*, Belgium: United Nations Educational, Scientific, and Cultural Organisation (Unesco).

Kakabadse, A. (1983) *The Politics of Management*, Aldershot: Gower Publishing.

Kolb, D. (1984) *Experimental Learning*, London: Prentice Hall.

Melton, K. and Tinsley, S.J. (1999) 'Outlook for greener marketing: unsettled and cyclonic', *Eco-Management and Auditing Journal* 6 (2): 86–97.

Pollack, S. (1995) *Improving Environmental Performance*, London: Routledge.

Schein, E.H. (1969) *Process Consultation: Its Role in Organisation Development*, California: Addison-Wesley.

Shelton, R.D. (1994) 'Hitting the green wall: why corporate programmes get stalled', *Corporate Environmental Strategy* 2 (2): 5–11.

Starkey, K. (1996) *How Organizations Learn*, London: International Thomson Business Press.

Tinsley, S.J. and Melton, K. (1997) 'Sustainable development and its effect on the marketing planning process', *Eco-Management and Auditing Journal* 4 (3): 116–26.

Welford, R. (1997) *Corporate Environmental Management: Culture and Organisations*, London: Earthscan Publications.

Welford, R. and Starkey, R. (1996) *Business and the Environment*, London: Earthscan Publications.

Index